H酒店+
Otel
Coastal Resort　滨海度假

1

佳图文化　主编

中国林业出版社

Contents 目录

■ **Information**　资讯

■ **Hotel Brand Management**　酒店品牌管理

008　Explore the Hotel Design in a Branding Context
探究品牌化背景下的酒店设计

■ **International Brand Hotel**　国际品牌酒店

014　W Singapore - Sentosa Cove
新加坡圣淘沙湾 W 酒店

042　Sheraton Shenzhou Peninsula Resort
神州半岛喜来登度假酒店

■ **New Hotel**　新酒店

062　Kempinski Hotel Haitang Bay Sanya
三亚海棠湾凯宾斯基酒店

■ **Special**　专题

076　The Naka Island, Phuket— A Luxury Collection Resort and Spa
普吉岛纳卡岛度假酒店——豪华精选度假胜地

098　Jumeirah Vittaveli
卓美亚维塔维丽度假酒店

116 Jumeirah Dhevanafushi
卓美亚德瓦纳芙希度假酒店

132 Sheraton Sanya Haitang Bay Resort
三亚海棠湾喜来登度假酒店

156 Sheraton Rhodes Resort
罗得岛喜来登度假酒店

174 The Ritz Carlton Hotel, Okinawa
冲绳丽思卡尔顿度假酒店

192 Angsana Laguna Phuket
普吉岛乐古浪悦椿度假村

210 Angsana Balaclava Mauritius
毛里求斯巴拉克拉瓦悦椿度假村

226 Sheraton Yantai Golden Beach Resort
烟台金沙滩喜来登度假酒店

240 Sheraton Bali Kuta Resort
巴厘岛库塔喜来登度假酒店

Resort Hotel　度假酒店

254 The St. Regis Bal Harbour Resort
巴尔港瑞吉度假酒店

270 Jumeirah Port Soller Hotel & Spa
卓美亚 Port Soller 水疗酒店

Information 资讯

希尔顿全球亚太地区第100家酒店开业

2013年5月2日希尔顿全球宣布,其在亚太地区的第100家酒店——曼谷素坤逸希尔顿逸林酒店盛大开幕。这家酒店的开业是希尔顿全球在亚太地区重要的里程碑事件。另外,希尔顿全球在亚太地区还有超过170家酒店,近50 000间客房正在筹备之中,目前的酒店品牌组合也预期会在未来五年内增长三倍。

The 100th Hilton Hotel in the Asia-Pacific Region Opened

On May 2nd, 2013, Hilton Hotels Group announced its 100th hotel in the Asia-Pacific region – DoubleTree by Hilton Hotel Sukhumvit Bangkok was opened. The operation of this hotel is important milepost event of Hilton International in the Asia-Pacific region. In addition, Hilton has more than 170 hotels in the Asia-Pacific region, 50000 rooms are being prepared, and the present hotel brand portfolios are expected to grow three times in the next five years.

洲际酒店集团落户金东 成金华首家五星酒店

4月19日,全球客房拥有量最大的洲际酒店管理集团,正式携旗下皇冠假日酒店签约金华最大的城市综合体——紫金湾。据介绍,洲际集团旗下酒店覆盖100多个国家和地区,拥有65万间客房,每年客流量超过1.4亿人次。入驻紫金湾的洲际皇冠假日酒店,是浙中唯一一家洲际酒店管理集团旗下的五星级酒店品牌。

InterContinental Hotels Group Settled Its First Five-Star Hotel in Jinhua

On April 19th, InterContinental Hotels Group which holds the largest number of guestrooms officially carried its Crowne Plaza Hotel to sign the largest city complex in Jinhua – Zijin bay. According to its introduction, InterContinental Hotels Group spreads its hotels in more than 100 countries and regions, with 650,000 guestrooms and guest flow of over 140,000,000 every year. The Crowne Plaza Hotel is the only five-star hotel brand of Intercontinental Hotels Group in Zijin Bay.

卡尔森旗下丽笙五星级酒店签约郑州天地湾

汉唐墅院天地湾,立足全球视野,斥资近8亿元倾力打造天地丽笙五星级国际酒店,由美国卡尔森环球酒店运营,于4月26日正式签约,改变了郑州北区没有五星级酒店的历史。天地湾将全能配套与郊区静谧完美结合,打造真正的高尚人居;以五星酒店、大型商业、优质教育全能配套,打造北区墅院生活中心;以核聚一城巅峰的墅区生活,革新北区墅院格局。

The Five-Star Radisson Blu Hotel of Carlson Settled in Zhengzhou Tiandi Bay

Hantang Villas Tiandi Bay is based on the global vision and spends nearly 800,000,000 Yuan to build the five-star Radisson Blu Tiandi Hotel which is operated by Carlson Hotels Worldwide; Radisson Blu Tiandi Hotel made an official sign on April 26th, which changed the history of Zhengzhou without a five-star hotel. Tiandi Bay perfectly combines its all-round supporting facilities with the quiet suburban area to create the indeed noble living environment. Five-star hotels, large commercial facility and high-quality education build the north villa community into a living center. Villa community concentrated in one city brings the North villa community pattern a new layout.

"中国的马尔代夫"云品酒店在云和正式开业

4月27日,由宋城集团重金打造的云品酒店,在丽水云和1929高原旅游度假区正式开业。酒店群以孤岛的形式凌水而建,宛若镶嵌在万顷碧波上的宝石,每间客房均以马尔代夫的总统套房样板精心设计,造价高达600万元,为目前世界上单间投资最高。云品酒店以其个性化的主题定位,独一无二的山水大观以及颠覆传统的酒店概念,与世界顶尖知名酒店接轨,将提升丽水旅游整体品质,开启度假型酒店新模式。

"Chinese Maldives" Yunpin Hotel opened in Yunhe

On April 27th, Yunpin Hotel, supported by Songcheng Group, was opened in Lishui Yunhe 1929 Plateau Tourism Resort. The hotel group is built along the water as an independent island, like a jewel set in the vast water surface; all rooms are designed with the standard of Maldives president suite with a cost up to 6,000,000 Yuan, which is currently the world's highest investment on single room. Yunpin Hotel has its personalized theme position, unique landscapes and anti-traditional hotel concept and perfectly integrates with the world top well-known hotels to will enhance the overall quality of Lishui tourism and open a new mode of holiday hotel.

三亚海棠湾民生威斯汀度假酒店7月开业

三亚海棠湾民生威斯汀度假酒店作为喜达屋酒店与度假村国际集团旗下威斯汀品牌中国大陆首家海滨度假酒店将于2013年7月隆重登场。据了解,由三亚民生旅业有限责任公司投资并打造的这家五星级酒店位于三亚海棠湾,紧邻即将开业的中国第一大免税购物商城。酒店共设有452间舒适客房、套房与别墅,每一间配备超大观景阳台。此外,该酒店还拥有独立的会议与多功能空间总面积达2 000 m²以上,包括豪华宴会厅与户外草坪。

Westin Sanya Haitang Bay Resort is to open in July

Westin Sanya Haitang Bay Resort, possessed by the Starwood Hotels and the International Resorts Wenstin brand, as well as the first Seaside Resort in China mainland, is to open in July, 2013. Invested and built by Sanya Minsheng Tourism Industry Co., Ltd., this five-star hotel is located in Sanya Haitang Bay, adjacent to the largest duty-free shopping mall in China. The hotel has 452 comfortable guestrooms, suites and villas, and each is equipped with super viewing balcony. In addition, the hotel also has an independent meeting room and functional space with a total area of more than 2,000 m^2, including luxury banquet halls and outdoor lawn.

青海将增9家五星级宾馆

随着青海知名度的不断提升,来青海旅游的外地人越来越多,"十二五"期间青海全省旅游将实现倍增目标,星级宾馆建设是实现倍增的重要内容之一。2013年,青海省新建、续建的四星级以上宾馆达28家,其中五星级9家。

Qinghai to Add 9 Five-Star Hotels

With the increasing popularity of Qinghai, more and more foreigners come there for tourism; during the period of "Twelfth Five-Year Plan" in Qinghai will achieve double goals on the tourism industry, and the construction of star-level hotels is one of the important multiplication contents to realize. In 2013, Qinghai has 28 new-built or continued hotels above four-star, of which there are 9 five-star ones.

美爵荣获年度最受喜爱国际五星级酒店品牌

全球领先的国际酒店集团及亚太地区最大的酒店运营商——雅高集团近日宣布,在中国旅游品牌总评榜上,美爵这一专为中国市场量身打造的高档品牌被评为"年度最受喜爱国际五星级酒店品牌"。中国旅游品牌总评榜由中国城市第一媒体旅游联盟主办,该联盟的成员是来自全国各地的多家主流平面媒体,通过独立、谨慎和公正的评判,评选出中国旅游行业内最具活力的航空公司、酒店运营商、旅行社和旅游局。

Grand Mercure Was Awarded the Most Popular International Five-Star Hotel Brand

The world's leading international hotel group and the largest hotel operator in Asia-Pacific region – Accor recently announced that on the Chinese Tourism Brand Billboard, Grand Mercure, a high-end brand specifically tailored for the Chinese market, was named "most popular International Five-Star Hotel Brand". Chinese Tourism Brand Billboard is sponsored by the China City First Media Tourism Alliance; the alliance members come from many main print medias all over the country, through independent, prudent and fair judgment, they will make a selection of the most vigorous airline company, hotel operator, travel bureau and tourism bureau in Chinese tourism industry.

2013年度英国最佳精品酒店设计

伦敦里程碑酒店,这所被《Condé Nast Traveler》杂志誉为2013年度英国最佳酒店的精品酒店,古典奢华,位于一座已被列为英国二级保护建筑的维多利亚宅邸内。酒店的每一个房间,豪华客房、奢华套房以及常住公寓,都经过精心的装饰布置。在色调丰富饱满的房间内,布满了定制的纺织品、古董家具和数量众多的艺术珍品。每一个房间都被赋予与众不同的主题风格,从艾伯特亲王到绅士俱乐部,从玛格丽特公主、马蒂斯与亨利·伍德,从摄政王到白金汉宫,犹如小型的英国历史博物馆。

The Best British Boutique Hotel Design in 2013

Milestone Hotel Kensington in London is named the annually Best British Boutique Hotel in 2013 by "Condé Nast Traveler"; it is classic and luxury, located in a Vitoria block that is listed as the two levels of protection of architecture. Every room, deluxe guestroom, luxury suites and resident apartments have been decorated meticulously. In the colorful room, there are many customized textiles, antique furniture and art treasures. Every room is given a subject style out of the ordinary, such as Prince Albert, gentleman's club, Princess Margaret, Matisse, Henry•Wood, Princeregent and Buckingham Palace, which just look like a small British history museum.

万豪国际一季度利润增长31%

酒店行业指标股万豪国际(MAR)一季度利润增长了31%,期间每可用酒店营收与客房价格继续强劲增长。万豪一季度盈利1.36亿美元,合每股收益43美分,去年同期盈利1.04亿美元,合每股收益30美分。营收增长了23%,至31.4亿美元,好于汤森路透调查的分析师的28.1亿美元预期。万豪预计二季度每股收益可达55至59美分,分析师的预期为55美分。该公司还将全年每股收益预期上调了3美分,至1.93至2.08美元。该股盘后上涨了1.1%,至42.95美元。该股今年迄今已上涨14%。

Marriott International First- Quarter Profit up by 31%

Marriott International Inc.'s (MAR) first-quarter earnings climbed 31% as the hotel industry bellwether continued to record stronger revenue per available room and higher average daily rates. Overall, Marriott reported a profit of $136 million, or 43 cents a share, up from $104 million, or 30 cents a share, a year earlier. Revenue jumped 23% to $3.14 billion, topping the $2.81 billion estimate from analysts polled by Thomson Reuters. In the second quarter, Marriott also predicted earnings of 55 cents to 59 cents a share, while analysts projected 55 cents. The company also raised its full-year earnings view by three cents to a range of $1.93 to $2.08 a share. Shares were up 1.1% to $42.95 after hours. The stock is up 14% since the start of the year.

Hotel Brand Management 酒店品牌管理

随着竞争的加剧，同类型同星级酒店的硬件日益接近，其产品的同质化日益增强，而品牌则是跨越这一障碍的撑杆。一件产品可以被竞争对手模仿，但品牌则是独一无二的，成功的品牌是持久的。因此，酒店要强化品牌个性来体现差异，突出竞争优势。中国本土的酒店管理公司也以各种方式在拓展自己的酒店品牌。理论界对于酒店品牌的研究由来已久，而酒店设计与品牌管理的关系问题则是其中颇具专业性的话题。

探究品牌化背景下的酒店设计

美国营销学家菲利普·科特勒认为品牌是一个名字、称谓、符号或设计，或是上述的总和，其目的是要使自己的产品或服务有别于其他竞争者。由于酒店产品与服务"不可触摸性"的特点，酒店品牌在市场营销中的作用越来越明显。经常旅行的人都会选择自己了解的和适合自己的酒店进行消费。这也是世界上品牌酒店所占比例越来越高的原因所在。

高速发展的中国酒店业，品牌已成为旅游者选择酒店的重要依据之一。如今，世界排名前10位的国际酒店管理公司均已进入中国市场，已有40多家国际酒店管理公司的60多个酒店品牌亮相中国，管理着500家以上的酒店，其增长势头依然强劲。

酒店品牌内涵与设计的兴起

酒店品牌有时代表了一个企业（集团）所属的所有酒店，有时只是其中之一。这里有两种情况：一是按照现代市场的操作模式，一个企业可以通过投资收购另一家企业，同时保留被收购的品牌；二是为了做不同的消费市场，在一个大的品牌下，开发出不同的酒店品牌以适应不同的顾客。

譬如人们熟悉的万豪国际酒店集团（Marriott）其旗下就拥有万豪（Marriott）、万丽（Renaissance）等10多个品牌，分别代表了不同的酒店概念。

酒店设计的形式始于20世纪80年代中期，源于世界著名设计师Phillippe Starck，他可以将牙刷变成艺术品，将电视变成玩具，其设计领域涉及建筑设计、工业设计、包装设计等，设计的产品从家具、灯具、高科技日用品，到服装、箱包、食品、汽车，应有尽有。其设计理念是：将一种文化概念的美丽升华为人文概念的美好。对环境和人文的尊重是Phillippe Starck的重要风格。法国总统密特朗曾请他设计过爱丽舍宫的内部，在日本设计的一系列风格独特的建筑使他成为表现主义建筑的代表还有巴黎高级艺术学院、波尔多机场控制塔等也都是他的杰作。

1990年，酒店业怪杰Ian Schrager请Phillippe Starck为纽约的派拉蒙酒店（Paramount Hotel）进行全面设计，由大堂的桌椅到房间的床柜到浴室牙刷，里里外外，全都出自Starck的手笔。客人住在派拉蒙酒店，就像住在Starck的设计产品陈列室一样。Starck的名气加上其风格独特的设计使派拉蒙酒店成为世界顶级的经典设计酒店。

寻求品牌与地域特征之间的平衡点

酒店设计是一门社会科学，对酒店建设及营运成本高低、投资与经营成功与否关系十分重大。由于一些投资者、设计者不重视，或不懂酒店的规划与设计，致使酒店设计中存在很多问题。

在酒店业发达国家，当一个新酒店项目启动之时，相关的机构包括业主、投资人、管理公司、设计单位等，都是同时进入的。实践证明这样的方式对酒当一个新酒店项目启动之时，业主、投资人、管理公司、设计单位等同时进入。实践证明这样的方式对酒店建设和经营十分重要，也是最理想的。设计者可以减少盲目设计，从而有更多的精力放在酒店品味和个性的营造上。

店建设和经营十分重要，也是最理想的。其益处在于：对业主、投资人而言，可以避免重复施工带来的经济损失；对管理公司而言是强化了品牌，便于日后有效的管理和经营；对设计单位（设计者）而言则是减少盲目设计，从而有更多的精力放在酒店品味和个性的营造上。

业主、投资人投资酒店，选择合适的品牌和酒店管理公司是与其自身条件、前期调研分不开的：品牌公司的扩展通常与其战略布局是一致的；设计师则必须在两者之间寻找平衡点，将品牌特征与地域特点有机的结合起来。而酒店概念规划要严格满足有关的法律法规、合理经济的满足酒店特定的经营需求、充分体现酒店的定位理念、将酒店的个性和周边的环境有机的融合。必须认识到酒店在经营过程中其形态和功能的不可变更性，及规划设计是影响酒店价值的重要因素。所以一般酒店设计公司都构架了具有国际视角、丰富的项目经验、多专业交叉的酒店概念规划设计顾问团队，将充分协调酒店管理和酒店设计之间的需要关系，前瞻性的策划拟建项目的概念规划方案。

酒店设计突出品牌文化传承

酒店设计不同于单纯的工业与民用建筑设计和规划，是包括酒店整体规划、单体建筑设计、室内装饰设计、酒店形象识别、酒店设备和用品顾问、酒店发展趋势研究等工作内容在内的专业体系。酒店设计的目的是为投资者和经营者实现持久利润服务，要实现经营利润，就需要通过满足客人的需求来实现。由于认识的局限，设计师常常将酒店看作是一个类型，而忽略消费者的差异和不同的酒店管理体系。对酒店品牌的认识可以帮助其深入和全面的理解酒店设计的本质特征和内在规律。

建筑是艺术，品牌酒店更因其形成过程所积淀的人文历史经典而展示无限的魅力。著名的凯悦(Hyatt)品牌开始并没有影响力，直到1967年凯悦在美国亚特兰大的新酒店开业，才使其名声大噪。这座由约翰·波特曼设计的宏伟建筑包含了一个庞大的中庭、透明的观光电梯和顶层的旋转餐厅，日后，这些都成为凯悦旗下的著名品牌君悦酒店的标志性设计，也成为其他酒店效仿的样本。如今走在世界各地的凯悦酒店，人们都能感受到这一品牌对建筑、对酒店环境的独特追求，因此也把它看作是凯悦的文化传承。君悦酒店除了它的独特设计还展示了它的豪华，如采用大理石和高档玻璃，以及颇具特色的照明等。

万豪酒店在视觉上更强调地域文化。在世界各地，它都能将当地的文化很好地融入自己的设计之中。在功能上则显示了它的细致入微，如为客人提供最大可能的舒适度，酒店功能布局的最大合理化，客房数量与公共区域的比例适当，酒店的流程合理化，为了适应更多的商务客人，不作过多的装饰。里兹·卡尔顿(RitzCarlton)酒店趋于传统，他们更多采用了木制品、座椅、沙发和老式花纹地毯，尽可能给人以舒适典雅的感觉，摆设也是具有历史内涵的物品。

香格里拉酒店的设计一向以清新的园林美景、富有浓厚亚洲文化气息的大堂特征闻名于世。香格里拉酒店大量分布于东亚和东南亚，即使开设在世界其他地区也有着同样的特点。与其类似的另一品牌是东方文华酒店集团。

一些著名的设计公司在承担酒店设计项目中除了保持自己独有传统的设计模式、设施特点外。更尊重所在国家地区的风土人情、人文历史，满足公众和旅游者的需求。酒店设计"地域化"的趋势对于全面理解品牌与设计的关系具有重要意义。

设计必须显示品牌特征

对于经常出入世界著名酒店的旅行者，往往不用提示就能猜出眼前酒店大致属于哪个品牌或类似酒店，原因就在于酒店的设计已经体现了它的管理模式、经营方式、接待礼仪等。当今酒店业品牌转换的情况时常发生，有时一家酒店刚开业不久就因为某种因素而转到另一品牌名下，装潢虽然很新，并处在同一等级，但接手的品牌公司还是会对它进行装修改造，或者至少进行重新布局。问题就在于不同酒店品牌的管理模式和经营理念存在差异。

20世纪90年代，在美国康涅迪克州的斯坦福，威斯汀酒店在接替经营才两个月遇到问题的TRUSTHOUSE FORTE饭店后，提出以贷款花200万美元改造大堂的条件接受管理合同，所用的钱将按经营损失来处理。问题就在于原饭店离威斯汀的标准差得很远。而威斯汀品牌是强调有形装置和高质量服务的。除了这些不足外，威斯汀热衷于追求新奇。威斯汀在签定管理合同前自己进行了可行性研究，预期性规划描绘了美好的前景。

对于威斯汀这个品牌，现在无论是在北京或是在上海，都能感受到类似的理念。20世纪末，上海先后有太平洋威斯汀变为太平洋喜来登酒店。波特曼香格里拉变为波特曼丽嘉酒店。为此，两家酒店都进行了大量的装修和内部调整，以适应各自的品牌特征。

消费群体决定酒店设计方向

消费群体决定了酒店设计的方向。比如旅游胜地的度假型酒店，在设计上主要是针对不同层次的旅游者，为其提供休息、餐饮和借以消除疲劳的健身康乐的现代生活场所。商务酒店一般具有良好的通讯条件，具备大型会议厅和宴会厅。以满足客人签约、会议、社交、宴请等商务需要。经济型酒店基本以客房为主，没有过多的公共经营区，实用大方。

每一酒店品牌都有明确的消费群体定位，如同属于喜达屋国际酒店集团(Starwood)的圣·瑞吉斯(St. Regis)i酒店是世界上最高档饭店的标志，代表着绝对私人的高水准服务。威斯汀(Westin)在酒店行业中一直位于领先者和创新者行列，它分布于重要的商业区，每一家饭店的建筑风格和内部陈设都别具特色。至尊精选(The Luxury Collection)是为上层客人提供独出心裁服务的饭店和度假村的独特组合，如华丽的装饰、壮观的摆设、先进的便利用具和设施等。W饭店(W Hotel)针对商务客人的特点对服务设施和服务方式、内容上有全新的设计。

酒店设计"地域化"的趋势对于全面理解品牌与设计的关系具有重要意义。酒店设计必须尊重所在国家地区的风土人情、人文历史，满足公众和旅游者的需要。

| Hotel Brand Management 酒店品牌管理

As competition intensifies, the facilities and equipment of the same star-rated hotel of the same type are becoming more and more similar. However, brand is an exception. A product can be imitated by competitors, but the brand is unique and a successful brand is long-lasting. Therefore, hotels have to strengthen the brand personality to reflect differences and highlight the competitive advantages. China's domestic management company should expand their own hotel brand in various ways. Theoretical cycle got researches on hotel brand for a very long time, and the relationship between hotel design and brand management is one of the most professional topics.

Explore the Hotel Design in a Branding Context

American marketing master Philip Kotler believes that brand is a name, a title, a symbol or a design, or to sum it up, its purpose is to make their own products or services that are different from those of other competitors. Due to the "can't be touched" characteristic of hotel products and services, hotel brand plays a more and more important role in marketing management. Frequent travelers always choose the hotel they understand and the hotel that suits them, that's also the reason why the world's brand hotels is in a growing percentage.

In the fast-developing Chinese hotel industry, the brand has become one of the most important elements for tourists to make their decisions. Today, there are more than 500 hotels of 60+ hotel brands under the management of more than 40 international hotel management companies in China, still having strong growth momentum.

The Rise of Hotel Brand Connotation and Design

Hotel brand sometimes represents all the hotels of an Enterprise (Group), and sometimes just one of them. There are two situations: first, in accordance with the operation mode of the modern market, an enterprise can buy another enterprise by investing, while retaining the original brand; second, in order to develop different markets, different hotel brands are established under a big brand to suit different customers.

For example the well-known Marriott Group owns more than 10 brands like Marriott and Renaissance, etc. which represent different hotel concepts.

The form of hotel design began in the mid-1980s, from the world-renowned designer Phillippe Starck, who can turn toothbrush into artwork and TV into toy, and he involves in architecture design, industrial design and packaging design, etc. The products he designed include furniture, lamps, high-tech daily necessities, clothing, bags, food, cars, everything. His design philosophy is: We have to replace beauty, which is a cultural concept, with goodness, which is a humanist concept. Respect for the environment and human is the important style of Phillippe Starck. French President Francois Mitterrand had asked him to design the interior of the Elysee Palace. A series of unique architectural style designed in Japan made him a representative of the expressionist architecture. In addition, Collège de Paris and Bordeaux Airport Control Tower are also his masterpieces.

In 1990, Phillippe Starck was invited by entrepreneur Ian Schrager to redesign the Paramount Hotel. From tables & chair in the lobby to cabinets to the toothbrush in the bathroom, inside and out, all of them are designed by Phillippe Starck. Staying at the Paramount Hotel is just like living in Starck's showroom. His fame and unique design style made Paramount Hotel the world's top design hotel.

Seeking Balance between Brand and Regional Feature

Hotel design as a subject of social science, matters to the cost of hotel construction and operation, the success of investment and operation. A large number of problems could be found in hotel design owing to investors and designers' disrespect or ignorance of hotel planning and design.

In countries with developed hotel industry, relative parties simultaneously participate in a hotel project in the starting stage, including proprietors, investors, management companies and design studios etc, which has been proved to be the most significant and ideal way for hotel construction and operation with benefits as following: to proprietors and investors, it could avoid economic losses due to repeated construction; to management companies, it strengthens the brands, facilitating the effective management and operation of future; to design studios(designers), it could reduce blind design and focus more on the construction of hotel taste and individuality.

Proprietors and investors' decision on hotel investment, hotel brand

> Relative parties simultaneously participate in a hotel project in the starting stage, including proprietors, investors, management companies and design studios etc, which has been proved to be the most significant and ideal way for hotel construction and operation. It will reduce designers' blind design and focus more on the construction of hotel taste and individuality.

and management companies is related to their self conditions and the prophase investigation; the expansion of brands is normally consistent with the strategic layout; designers are required to seek the balance point between both, and to organically connect the brand characteristic to regional features. Hotel conceptual planning is expected to strictly satisfy requirements of relevant laws and regulations, reasonably and economically meet hotel-specific operating demands, fully reflect the orientation concept of hotel, and integrate the hotel individuality with surrounding environment. It is significant to see the non-alterability of form and function in hotel operation, and planning design as the essential factor of hotel value. Thus hotel design companies are generally qualified for their international perspective, rich project experience, multi- interdisciplinary consultant teams of hotel conceptual planning

Hotel Brand Management 酒店品牌管理

> The trend of localization in hotel design is significant for comprehensive understanding of the relation between brand and design. Hotel design shall pay respect for the national and regional customs, culture and history, satisfying the demands of public and visitors.

and design. They will coordinate the demand relation between hotel management and hotel design, and present conceptual planning of the proposed project with forward looking.

Hotel Design Highlighting Brand Culture Transmission

Different from planning and design of industrial and civil architectures, hotel design is a professional system composed of overall planning, design of unit building, interior design, hotel image design, consultant on hotel equipments and supplies, study of hotel development trend etc. Hotel design aims at bringing lasting profit to investors and proprietors, and it is realized through meeting the needs of customers. Some designers view hotel as a single type in disregard of customer diversity and different management systems, due to limit of understanding. The realization of hotel brand will help them further and comprehensively understand the essential characteristics and inherent law of hotel design.

Architecture is an art and band hotel shows its infinite charming from the humane and historical classicism accumulated during its establishment. The renowned brand Hyatt was not eminent until the opening of its new hotel in Atlanta, USA in 1967. The grand building designed by John Portman contains a large atrium, transparent sightseeing elevators and revolving restaurant on the top floor, all of which becomes the iconic design of Grand Hyatt hotels as well as the imitation sample of other hotels afterwards. Hyatt hotels all around the world reflect their unique pursuit of architectures and hotel environments, which could be viewed as the cultural transmission of Hyatt. Grand Hyatt hotels show not only their unique design but also their luxury, like marble, upscale glass and distinctive lighting etc.

Marriott hotels pay more attention to regional culture in sense of vision. They are successful in the integration of regional culture in their design. They functionally show their minute and delicate service, i.e. offering customers the best comfort as they can: maximum rationalization of hotel functional layout, appropriate ratio between guest room quantity and public area, rational process and no excessive decoration to service more business guests. Ritz Carlton hotels tend to be traditional, using woodworks, chairs, sofa and carpets in vintage pattern to provide comfortable and elegant sense. Their furnishings are with historical connotation.

Shangri-La hotels are noted for their fresh garden landscape and lobbies with great Asian culture. Shangri-La hotels are massively located in East Asia and Southeast Asia, and the hotels located in the rest of the world share the same features. Another brand similar to Shangri-La is Mandarin Oriental Hotel Group.

In addition to maintain the own unique and traditional design patterns, facility features, some famous design companies also pay their respect for the national and regional customs, culture and history, satisfying the demands of public and visitors. The trend of localization in hotel design is significant for comprehensive understanding of the relation between brand and design.

The Design Should Show Brand Attributes

For travelers frequently in and out of the world famous hotels, usually, they don't get a tip but still can generally puzzle out which brand the hotel belonging to, because the design of the hotel has reflected its management mode, operation mode and reception manners, etc. Nowadays, the transformation of hotel industry brand happens frequently, sometimes a hotel just starts operating but soon comes to be transferred to another brand because of some factors; although its decoration is new and at the same level, but the took-over brand company will make a renovation of it, or at least to turn the layout. Problem is that different brand hotels have different management model and business philosophy.

In the 1990s in Stanford of Connecticut, US, when Westin Hotel took over the TRUSTHOUSE FORTE, a hotel closed down only two months after its opening, it proposed that the lenders should pay $2 million to reform the lobby as the conditions of the management contract, and the money will be treated as the operating loss. Problem is that the original hotel deviated a lot from the Westin Hotel standard. While Westin brand emphasized on physical device and high quality service. In addition to these deficiencies, the Westin is keen on the pursuit of novelty. Westin made a feasible study before signing the management contract to plan a prospective and wonderful prospect.

Now whether in Beijing or Shanghai, people can feel a similar idea of Westin. In the end of the 20th century, the Pacific Westin was transformed to Pacific Sheraton Hotel in Shanghai, and Portman Shangri-la changed to the Portman Hotel. These two hotels both went on a lot of decorative and internal adjustment so as to adapt to their own brand characteristics.

Consumer Group Determines Hotel Design

Consumer group determines the direction of the hotel design. The design of holiday hotels in the tourist destination is mainly aimed at different levels of tourists, providing them rest area, food, beverage and modern fitness and recreation facilities to eliminate the tourists' fatigue. Business hotel generally has good communication conditions, which has large conference hall and banquet hall to meet the guests' business needs of conducting a sign, meeting, social activities and banquets. Economy hotels give first place the guestrooms without too many public areas, practical and tasteful.

Every hotel brand has its clear target consumer group; for example, St•Regis Hotel, belonging to Starwood International Hotel group, stands for the world's most upscale hotel, representing the absolute high-level personal service. Westin has been leader and innovator in the hotel industry, distributing in the important business districts, and each hotel has unique architectural style and interior furnishings. The Luxury Collection is a unique combination of hotels and resorts provided for the upper guests, such as luxuriant adornment, magnificent decoration, advanced facilities, equipment and facilities, etc. According to the characteristics of the guests, W Hotel goes on brand-new design on the service facilities, service mode and service contents.

W Singapore - Sentosa Cove | 新加坡圣淘沙湾 W 酒店

Keywords 关键词
Personality and Fashion 个性时尚
Sensory Design 感官设计
Geographic Feature 地域特色

品牌链接

作为喜达屋酒店与度假村国际集团旗下酒店，W 酒店是圣·瑞吉斯、威斯汀、喜来登等豪华酒店的姊妹品牌。每家 W 酒店的独特设计灵感都来自它的所在地，将当地影响力融入到最尖端最前沿的设计之中，创造出了白天尽情玩耍和工作，夜晚释放能量的美妙之地。所以，虽然都是 W 酒店旗下的酒店，但是每一间都有它自己的风格，都被酒店的合作者赋予了新的生命，而合作者们都是设计、音乐和时尚新闻领域的尖端人物。

About W Hotels

The brand W hotel is a subsidiary of Starwood Hotels which also operates Westin, Sheraton, St. Regis and a few smaller niche brands. Each hotel and resort is uniquely inspired by its destination, mixing cutting-edge design with local influences and creating a place to play or work by day or to mix and mingle in high-energy spaces by night. So, while every hotel is unmistakably a W hotel, each has its own personality that is brought to life by our collaborators who are always on the cusp of what's new and next in design, music and fashion.

酒店地址：新加坡新加坡市海洋大道 21 号
电　　话：(65) 6808 7288

Address: 21 Ocean Way, Singapore 098374, Singapore
Tel: (65) 6808 7288

| International Brand Hotel 国际品牌酒店

品牌发展历程

W 酒店将独立酒店的个性化时尚风格与主流商务酒店值得信赖、始终如一的卓越品质和体贴周到的服务融为一体，从而重新诠释了以当代设计为主导打造奢华现代生活方式的住宿体验。

第一家 W 酒店——纽约 W 酒店（位于 49th St. 和 Lexington 交界处）于 1998 年 12 月正式开业。之后不久，W 酒店便成为业界瞩目的焦点，在短短两年间便创下了在洛杉矶、芝加哥、西雅图和首尔等地开设 12 家分店的骄人成绩。

每一家 W 酒店均设有特色餐厅及酒吧，不仅吸引了大批酒店住客，也倍受当地居民的青睐。通过与著名厨师 Drew Nieporent 的合作，纽约首家 W 酒店的 Heartbeat 餐厅以及西雅图 W 酒店的 Earth & Ocean 餐厅陆续取得空前成功。纽约联合广场 W 酒店 Olives 餐厅的 Todd English 以及达拉斯维多利亚 W 酒店 Craft 餐厅的 Tom Colicchio 等厨师也纷纷推出了其自创的精品佳肴。

全球的每一间酒店和度假村均综合了酒店诞生地（纽约市）的积极的、有活力的、思想超前的态度，成就了完全适合其所在地方的氛围。这是一个反映 W 酒店愿景的方法，一个混合了各种氛围和关键素材，创造出集包容和社会性于一身的场所。

History & Development

Combining the personality and style of an independent hotel with the reliability, consistency and attentive service of a major business hotel, W Hotels has redefined the luxury and design-led lifestyle hotel experience.

The W Hotels experience began with the W New York (49th St. and Lexington Ave.), which opened its doors in December 1998. An instant phenomenon, its success drove the development of more than a dozen new properties in colorful destinations – including Los Angeles, Chicago, Seattle and Seoul – in an unprecedented two-year span.

Each hotel offers signature restaurant and bar areas that attract not only hotel guests, but local tastemakers as well. A collaboration with renowned chef Drew Nieporent led to the success of Heartbeat, the restaurant in the first New York location, followed by Earth & Ocean at W Seattle. The creative excellence of chefs like Todd English of Olives at W New York – Union Square and Tom Colicchio of Craft at W Dallas – Victory extended W Hotel's foray into the culinary world.

The approach W Hotel started with in 1998 still energizes itself today. Each hotel and retreat worldwide synthesizes its birthplace's (New York City) energetic, vibrant, forward-thinking attitude into an atmosphere entirely appropriate to its destination. It's an approach that reflects the W Hotels vision of the hotel as a mix of vibes and elements that mingle to create a welcoming and social gathering place.

International Brand Hotel 国际品牌酒店

酒店概况

新加坡圣淘沙湾 W 酒店坐落于东西方文化的荟萃之地——圣淘沙岛上。该岛地处新加坡南端，拥有充满活力的海滩酒吧、时尚别致的美食餐厅、放松舒缓的水疗馆以及众多美丽迷人的旅游景点，如环球影城和位于圣淘沙名胜世界的娱乐场。

Overview

East meets West on the island of Sentosa, where W Singapore – Sentosa Cove resides. Situated just off the southern tip of Singapore, the island's vibrant beach bars alternate with chic restaurants, indulgent spas, and tempting attractions, like Universal Studios or the casino at Resorts World Sentosa.

Hotel+_019

International Brand Hotel 国际品牌酒店

酒店特色

在酒店的私人泊船区靠岸，可前往 AWAY® 水疗中心享受舒缓身心的护理，或到设备完善的健身房健身，或纵身跃进 WET® 室外泳池畅游一番，或在回覆式花园露台上举办迎宾活动，或休闲放松，亦或在 W 酒廊的惬意氛围中品尝创意鸡尾酒并观看 DJ 现场表演。酒店的连线商务中心可让顾客从始至终保持连通，而随时 / 随需® 服务则全天候满足顾客的一切需求。W 精品店出售独具 W 酒店特色的精美商品，宾客可挑选一些作纪念。

Feature

Pull up to the private docking berths, and then relax with soothing treatments at AWAY Spa. Energize in the fully equipped GYM or with a splash at WET, the outdoor pool. Or there's the laidback Garden Terrace—for pre-function events or general relaxation—and W Lounge, which offers a relaxing space by day and innovative cocktails and a live DJ by night. Throughout it all, the WIRED Business Center keeps the guest connected while Whatever/Whenever attends to everything, and anything, else. And when it comes time to take off, bring a piece of W with you, thanks to the signature W Hotels – The Store.

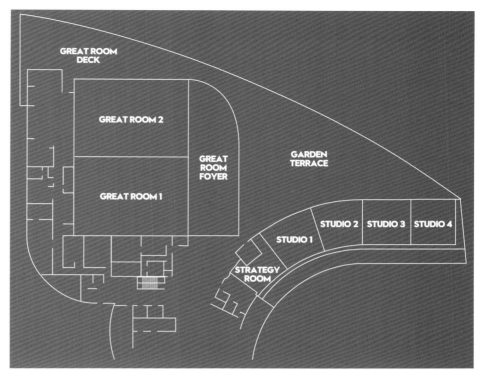

Ballroom & Pre-Function

International Brand Hotel 国际品牌酒店

酒店配套

■ 随时/随需·服务

新加坡圣淘沙湾W酒店时刻将客人的愉悦体验放在首位。因此，酒店提供随时/随需·服务。无论客人需要什么，无论何时需要，绝对没问题！

■ 厨桌餐厅

休闲的厨桌餐厅设有现场美食秀，供应各种现代的印度、中国、日本及国际美食，包括寿司、点心、印度烤饼、炭烤披萨及甜点等。特色餐厅从早上一直营业至晚上，设有精致装潢、时尚天然的装饰以及回覆式室内外布置，是朋友欢聚或举办难忘庆典与晚宴的理想之选。

互动式美食秀提供一系列国际美食，从中式点心、泰式面条到日本寿司、炭烤披萨和印度铁烤饼等，不一而足。"随心选择"零点菜单提供各种当地特色美食（如马来椰浆饭）和创意西餐（如烘焙干酪馅饼与酒香雪梨配芝麻菜、甜核桃与三块布里干酪），更有各种美味甜点与绝佳佐餐饮品打造不可比拟的美妙体验。

Services & Amenities

■ Whatever/Whenever® Service

At W Singapore – Sentosa Cove, your pleasure is the priority. That's why they offer the Whatever/Whenever® service. Whatever you want, whenever you want it. No problem.

■ The Kitchen Table

Sample a remix of contemporary Indian, Chinese, Japanese and international flavors at the kitchen table, where live cooking stations—including sushi, dim sum, tandoori, wood-fired pizza, and dessert options—fuse with casual surroundings. Open daily from morning till night, the signature restaurant welcomes you with inviting décor, combining sleek, natural surfaces with nature-inspired details, and a laidback indoor-outdoor vibe that encourages mixing and mingling, celebrations and memorable meals.

Explore the extensive selection of international flavors at our interactive series of live cooking stations serving everything from Chinese dim sum and Thai noodles to Japanese sushi, wood-fired pizza and Indian tandoori. And of course there is always the "Explore Your Options" à la carte menu, featuring local specialties like Nasi Lemak and innovative Western flavors, like a Baked Gruyere Tart and Barolo Spiced Pears with Arugula, Candied Walnuts and Triple Brie. To top it all off, try the tempting desserts and perfectly paired beverages.

International Brand Hotel 国际品牌酒店

■ AWAY® 水疗中心

AWAY® 水疗中心，木制雕刻与舒缓的天然表面营造出休闲的雨林环境，让人远离尘嚣，沉浸在静谧的氛围中。Detox 更衣区设有电子储物柜，安全存储私人物品。还设有桑拿室、蒸汽浴室以及带身体喷流的活力泳池舒压区。Immerge 香草雨、Aqua 实验性淋浴以及著名的薇姿淋浴均带来焕发活力的休闲体验。此外，在 112 m² 的私人护理室或共振室可享受各种按摩、面膜、身体及美容护理、美甲美趾、打蜡等服务（使用专属的 ILA – 水疗产品或 OPI 美甲产品）。

■ AWAY® SPA

Escape to the serene AWAY Spa, where relaxing rainforest-like surroundings feature wooden sculptures and soothing natural surfaces. Unwind in the Detox changing area, where electronic lockers keep personal items safe. Then head to the decompression area, home to a sauna, steam room and vitality pool with body jets. Immerge, an herbal bath; Aqua, an experiential shower; and famous Vichy showers add to the revitalizing experience.

Settle into the hotel's private 112-square-meter treatment room or resonance room and select from the innovative array of massages, facials, body and beauty treatments, manicures and pedicures, waxing and more performed with exclusive ILA-Spa products or OPI nail products.

International Brand Hotel 国际品牌酒店

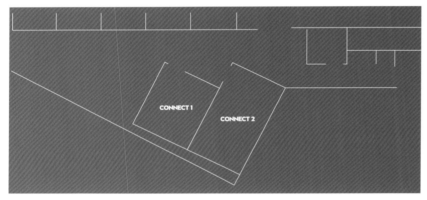

Meeting Rooms

■ W 酒廊

这里看上去像一个活力四射的传统大堂，设计概念是融休闲与欢庆于一体，体现了独具特色的 W 能量与新加坡圣淘沙湾 W 酒店的绝妙设计。白天，可在此免费使用 Wi-Fi 接入，或与朋友欢聚，酒廊还提供桌游和杂志等。日暮时分，这里就会幻化成 WOOBAR 酒吧。啜饮混合莫吉托，品尝美味简餐，还可一尝充当现场 DJ 的感觉。

■ W Lounge

The vibrant takes on a traditional lobby, this conceptual plan unites relaxation and celebration—both heightened by signature W energy and W Singapore - Sentosa Cove's electrified design. Tap into the complimentary Wi-Fi by day, or simply hang out with friends in the laidback space. W LOUNGE has board games, magazines and more. As the sun sets, the room revolutionizes into WOOBAR. Sip a remixed mojito. Taste epicurean light bites. And mingle as a live DJ amps up the soundtrack.

International Brand Hotel 国际品牌酒店

■ 新加坡圣淘沙湾 W 公寓

新加坡圣淘沙湾 W 公寓坐落于新加坡最负盛名且最为富裕的码头住宅小区——圣淘沙湾水滨住宅小区，开辟了一片"远离尘世喧嚣的世外桃源"，让人尽享超凡脱俗的生活方式。公寓共设有228套超豪华公寓，其中包括双卧室、四卧室和顶楼套房。每一套 W 公寓都代表着时尚风格的又一新篇章，已被打造成为对现代生活独特而个性化的重新诠释。富有艺术气息的建筑处处流露出舒适、奢华与创新，而新与旧、本地与全球风格更在此交相辉映，带来赏心悦目的绝妙奇迹。W 公寓设计注重本地特色，旨在打造与周边环境相得益彰的独特平衡。

■ The Residences at W Singapore – Sentosa Cove

Nestled amidst the exclusive waterfront residential enclave of Sentosa Cove—Singapore's most prestigious and affluent marina residential community, The Residences at W Singapore – Sentosa Cove is designed to offer an extraordinary lifestyle experience—an "escape within an escape". The Residences at W Singapore – Sentosa Cove will feature 228 ultra-luxurious residences, ranging from two to four bedrooms along with penthouses. Every W Residence is a new chapter in a storybook encounter of style, and the hotel has created its Residences to be unique and individual expressions of modern living. The artistry of our architecture coupled with the comfort, luxury and whimsy within, influences of old and new, local and global, come together in playful wonder. W Residences are designed to accentuate qualities indigenous to the locations, creating a unique, balanced relationship with the environment.

International Brand Hotel 国际品牌酒店

■ WET® 室外泳池

WET® 室外泳池面积达 1 338 m²，可饱览中国南海波光粼粼的壮阔美景，并免费提供带 iPad、iPod 及喷雾风扇的私人凉亭，是客人在白天或晚上享受清凉时光的绝佳之选。

■ W 精品店

W 精品店出售当地有代表性的纪念品、必备品、顶尖旅行配件及 Bliss Spa® 水疗美容产品。精品店位于酒店二层，所售商品从当地设计师的最新作品到酒店珠宝系列 Surrender to Vice and Vanity 再到新加坡圣淘沙湾 W 酒店的创意礼品，应有尽有。

■ 健身中心

健身中心面积达 235 m²，设有先进的健身器材，可通过落地窗观赏到迷人的花园景观。先进的泰诺健有氧健身器材配有个人电视，可轻松为健康充电，也可选择力量训练设备及其他设备进行健身。酒店还提供瑜伽垫、健身球与健身实心球、吊袋、Fitvibe 与 Kinesis One 设施等。并免费提供瓶装水、毛巾及带一次性保护套的耳机等。

■ WET®Outdoor Pool

Cool off day or night at WET®, The 1,338-square-meter outdoor pool with views of the sparkling South China Sea and complimentary private cabanas featuring an iPad, iPod and mist fan.

■ W Hotels - The Store

Pick up a local keepsake, a wardrobe staple, cutting-edge travel accessory or covetable Bliss Spa® beauty product at W Hotels - The Store. Situated on the second floor of the resort, it is stocked with everything from the latest offerings from local designer Mae Pang to a resort collection by Surrender to Vice and Vanity jewelry to an expertly curated selection of W Singapore - Sentosa Cove original gifts.

■ GYM

Stay in shape your way in the 235-square-meter, state-of-the-art fitness facility, GYM, with panoramic garden views through floor-to-ceiling windows. Get moving on top-of-the-line Technogym cardio machines with individual televisions before using the strength machines and alternative equipment. There are yoga mats, fit and medicine balls, punching bags and more, along with Fitvibe and Kinesis One technology. And don't forget to help yourself to the complimentary bottled water, towels and headphones with disposable covers.

International Brand Hotel 国际品牌酒店

■ **客房**

新加坡圣淘沙湾 W 酒店设有各种绝妙的客房与套房，均配备精美装饰、情境照明、先进科技与梦幻设备，让每一次入住均至臻完美。馋猫宝罐带来可口小食，而客房内餐饮服务则全天供应美食。迷你吧、法式压滤咖啡机及热水壶则随时准备好饮品。

客房内娱乐系统包括 iPod 基座、40 英寸三星 LED 电视及 DVD 播放器，带来无与伦比的休闲体验。Wi-Fi 和高速上网接入随时保持连通。热带雨林淋浴间或独立浴缸可享受专属 Bliss® 卫浴六件套带来的极致沐浴体验。然后穿上舒适浴袍与拖鞋，走到独具特色的 W 睡床上，尽享由 350 纱织密度床单、被子与 W 枕头带来的酣眠体验。

■ **连线商务中心**

连线商务中心配有两个带高速上网接入的电脑工作站与装订及覆膜机，并提供包装与运输服务，即使远离常规办公室，也能照常轻松办公。

■ Rooms

Revel in the exhilarating guest rooms and suites of W Singapore – Sentosa Cove, where vibrant décor, mood lighting, state-of-the-art technology and visionary amenities enhance every stay. Indulge in a snack from the Munchie Box or our around-the-clock in-room dining service. Or sip from the mini bar, French press coffee maker or hot water kettle.

Amplify downtime with in-room entertainment, ranging from an iPod docking station to a 40-inch Samsung LED TV and DVD player. And connect via Wi-Fi and High Speed Internet Access. Finally, recharge in the rainforest shower or separate bathtub with exclusive Bliss®Spa sinkside six bath amenities before wrapping up in a robe and slippers and retreating to signature W pillow-top beds, where 350-thread-count sheets, a down comforter and W pillow menu promise a rejuvenating night's sleep.

■ Wired Business Center

Amply outfitted with two workstations with High Speed Internet Access, packing and shipping services, plus a binding and laminating machine, WIRED Business Center is your office away from the office.

International Brand Hotel 国际品牌酒店

Sheraton Shenzhou Peninsula Resort | 神州半岛喜来登度假酒店

Keywords 关键词
- Coastal Resort 休闲海湾
- Spa 水疗护理
- Modern Style 现代风格
- Functional Diversity 功能多样

酒店地址：中国海南万宁神州半岛旅游度假区
电　话：(86)(898) 6253 8868

Address: Shenzhou Peninsula Resort District, Wanning, Hainan
Tel: (86)(898) 6253 8868

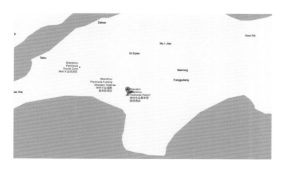

品牌链接

喜来登酒店与度假村集团（Sheraton Hotels and Resorts）是喜达屋（Starwood）酒店集团中最大的连锁旅馆品牌，而它也是集团中第二老的酒店品牌（最老牌的是威斯汀）。喜来登的酒店型态有许多种，从一般的商业旅馆到大型度假村都有；喜来登品牌一直力图维持高品质形象，在世界上的喜来登酒店有超过一半被当地机关评选为五星级酒店。喜来登酒店据点分布极广，遍布五大洲，从香港到斯里兰卡到埃及及津巴布韦等国都可见其旅馆。喜来登总部在美国纽约的白原市。

喜达屋酒店都有良好的选址，主要分布在大城市和度假区。集团酒店选址的标准是：所在区域的发展史表明，该地区对提供全方位服务的豪华高档酒店有大量、持续增长的需求。作为酒店业豪华高档细分市场中最大的酒店集团，喜达屋酒店的规模有力地支持它的核心市场营销和预定系统。喜达屋酒店在把重点放在豪华高档细分市场同时，其各种品牌分别侧重于该市场中不同的二级市场。喜达屋酒店在赌场业也占据着重要的位置，它主要是通过Caesars品牌来经营此业务。喜达屋酒店为休闲度假旅游者提供着宾至如归（home-away-from-home）的服务。

About Sheraton Hotels and Resorts

Sheraton Hotels and Resorts is Starwood Hotels and Resorts Worldwide's largest and second oldest brand (Westin being the oldest). It occupies many types of hotel properties, ranging from general commercial hotels to large-scale resorts with high-quality image; more than half of Sheraton hotel worldwide are recognized as five-star hotels, with a widespread distribution across five continents i.e. Hong Kong, Sri Lanka, Zimbabwe etc. It headquarters in White Plains, New York.

Starwood hotels are located mainly in big cities or resort districts, in accordance with the site selection standards of group hotel that the selected districts own an increasing and great demand for full-serviced luxurious high-end hotels. As the greatest hotel group in the luxurious high-end market of hotel industry, Starwood holds its coral market promotion and booking system supported by its scale. It lays its emphasis on luxurious high-end market and its brands are targeted at the secondary markets. It also has a great achievement in casino industry, mainly represented by Caesars brand. Sheraton hotels address themselves to offer the visitors services of home-away-from-home.

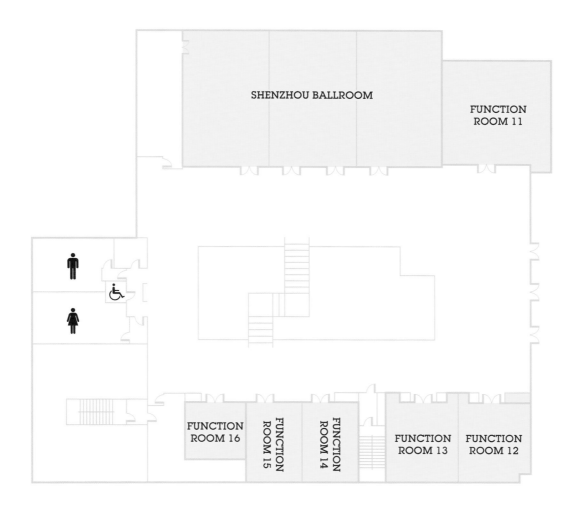

酒店概况

海南岛地处中国南海,是中国南部沿海的一处天堂度假胜地,素有"东方夏威夷"之美誉。 神州半岛喜来登度假酒店位于海南岛东海岸的万宁市,坐拥葱郁繁茂的棕榈树、蔚蓝无垠的海洋和气势恢宏的群山。酒店坐落于风景如画的综合度假区内,可欣赏到美丽壮观的海景,并拥有 8 km 长的白色沙滩和休闲海湾。酒店距三亚凤凰国际机场(SYX)仅 100 km 之遥。

Overview

Situated in the South China Sea, just off China's southern coast, Hainan Island is renowned as the "Hawaii of the East." Surrounded by palm trees, azure water, and picturesque mountains, Sheraton Shenzhou Peninsula Resort is located in Wanning, on east coast of Hainan. Overlooking the sea and set amid an integrated community, the resort boasts eight kilometres of white sand beaches and relaxing coves. It is 100 km away from Sanya Phoenix International Airport (SYX).

International Brand Hotel 国际品牌酒店

International Brand Hotel 国际品牌酒店

酒店特色

大堂网络中心——随心连动@喜来登（体验在微软®）轻松工作、社交、品尝小吃并与亲友保持联系。在另一地点，3-12岁的小客人们还可与新朋友一同体验各种创意活动、趣味游戏及艺术和手工艺活动。Core® Performance 喜来登健身中心可提供私人训练指导和最完美的健身体验。此外，客人还可在迷人热带花园中的泳池内尽情畅游，在喜来登炫逸水疗中心彻底放松，在温馨奢华的氛围中尽享独家设计的水疗护理服务。

Feature

Link@Sheraton Experienced with Microsoft® is the lobby connectivity hub, available for working, relaxation, socialization, grabbing a snack, and staying connected to friends and family. Elsewhere, young guests from 3 to 12 years old can enjoy creative activities, fun games, and arts and crafts with new friends. Sheraton Fitness Programmed by Core® Performance offers personal training expertise and best fitness experience. Customers can also take a dip in the swimming pool amid enchanting tropical gardens, relax completely at Shine Spa for Sheraton, which provides exclusively-designed spa treatments in a warm and upscale atmosphere.

International Brand Hotel 国际品牌酒店

酒店配套

■ 餐饮

酒店的多元化餐厅和酒吧提供丰盛可口的美食和佳酿。这一喜来登特色全天候餐厅以色彩亮丽的挂毯装饰其间，所有菜品均采用来自当地市场的最新鲜食材，并设有开放式厨房，以强化互动性餐饮体验。露天大堂吧内简单放松，是品尝最爱饮品和点心、享受简单生活的理想之地。

Services and Amenities

■ Dining

The food and beverages of the diverse restaurants and bars with the tastes of the world at Feast are the foundations for enjoyable experiences with family and friends. Sheraton's signature all-day dining restaurant features a tapestry of bright décor, the freshest ingredients directly from local markets, and open kitchens that encourage an interactive dining experience. The open-air lobby lounge is the ideal place to enjoy the simpler things over your favorite beverage and pastry.

International Brand Hotel 国际品牌酒店

■ 会议与活动

神州半岛喜来登度假酒店拥有一系列可根据不同需要进行灵活配置的会议室和活动厅。室内活动空间面积共达 900 m²，其中包括一间可容纳 440 位宾客的优雅无柱式豪华宴会厅、六间会议室和一个宽敞迎宾区。所有会议室均配有无线上网接入和电话会议联网设施。

别具一格的婚礼教堂是举办婚礼仪式的理想之地，一流的海滨位置和宜人的气候让酒店的户外场所终年可用。神州半岛喜来登度假酒店拥有一支专业的 MICE（会议、奖励、大会和活动）策划专家团队，他们将在景色宜人的环境中以丰富广博的知识和独具创意的团队建设方案为宾客打造一场成功圆满的会议或活动。

■ Meetings & Events

Sheraton Shenzhou Peninsula Resort has a variety of meeting and event rooms that can be transformed to suit your needs. The indoor function space totals 900 square meters, which includes an elegant, pillarless grand ballroom accommodating up to 440 guests, six meeting rooms, and a spacious pre-function area. All rooms are equipped with wireless High Speed Internet Access and teleconference networking.

A ceremony in the unique wedding chapel is ideal for holding a special ceremony and thanks to our superb beachfront location and magnificent climate, outdoor venues are available year-round. Sheraton Shenzhou Peninsula Resort provides MICE (Meetings, Incentives, Conferences, and Events) planners who can provide guidance not only with their extensive knowledge but also with creative team-building programs in a picturesque setting.

International Brand Hotel 国际品牌酒店

International Brand Hotel 国际品牌酒店

■ **客房及套房**

神州半岛喜来登度假酒店拥有276间豪华客房（52 m²）和32间套房（104 m²或更大）。所有客房均装潢精美，并配有特色的喜来登甜梦之床（Sweet Sleeper™）、壁挂式液晶纯平电视、高速上网接入和现代化设施，可供商务和休闲旅客尽享舒适便利的住宿体验。所有客房均可将壮丽海景一览无余，步出超大私人阳台可尽情感受中国南海的温柔海风。高尔夫海景客房可欣赏到迷人的自然风光和美丽果岭，而全景客房和套房则可将山峦与大海的壮观全景尽收眼底。

■ Rooms and Suites

Sheraton Shenzhou Peninsula Resort comprises 276 deluxe rooms (52 square meters) and 32 suites (104 square meters or larger). All accommodations are tastefully furnished with the signature Sheraton Sweet Sleeper™ Bed, wall-mounted LCD Flat Screen Television, High Speed Internet Access, and modern facilities to comfort both leisure and business travelers. All rooms feature uninterrupted ocean views. Step onto the oversized private balcony and customers could enjoy the breeze from the South China Sea. Golf Ocean View Rooms overlook spectacular landscaping and beautiful greens, while Panorama Rooms and Suites provide full views of both mountains and sea.

International Brand Hotel 国际品牌酒店

New Hotel 新酒店

Kempinski Hotel Haitang Bay Sanya | 三亚海棠湾凯宾斯基酒店

Keywords 关键词

Li and Miao Minority Culture 黎苗文化

Tropical Temperament 热带风情

Comfortable and Elegant 舒适典雅

Classics Collection 古典收藏

酒店地址：中国海南省三亚市海棠湾海棠北路2号
电　话：+86（0）898 8865 5555

Address: No.2 Haitang North Road, Haitang Bay, Sanya, Hainan Province, China
Tel: +86(0) 898 8865 5555

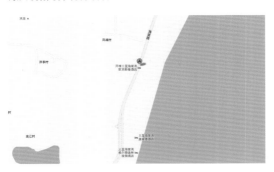

品牌链接

凯宾斯基酒店是世界上最古老的豪华酒店，最初建立于1897年。酒店集团创建于德国，现旗下酒店遍布欧洲、中东、非洲、南美洲和亚洲，在北京、柏林、布达佩斯、伊斯坦布尔、德累斯顿和圣莫里茨等地拥有45处以上的私人酒店和特色酒店。该酒店集团目前包含16处环境优美的度假胜地，每处都提供优越的整套休闲设施、豪华水浴和令人惊叹的地理位置。

凯宾斯基酒店重塑了"礼宾"这一理念，确保礼宾团队最大限度地重视每一位客人，不仅满足而且要超越客人的期望，从而始终获得高度推崇。为了体现客人的重要性，酒店开发出了一系列礼宾服务项目，包括商务礼宾、私人礼宾以及会议礼宾等。多年以来，凯宾斯基酒店已经发展成为富有创新性并受到高度推崇的豪华酒店集团，且因为能够满足和超越高品位国际游客的需求而闻名于世。每家酒店都提供优良的服务，且将酒店的特色和当地风格融入其中。

About Kempinski

Founded in Germany in 1897, Kempinski Hotel is the oldest luxury hotel in the world. Today, Kempinski has its group hotels all over Europe, Middle East, Africa, South America and Asia. In Beijing, Berlin, Budapest, Istanbul, Dresdon and St.Moritz, Kempinski also possesses over 45 private hotels and special hotels. Besides, Kempinski Group also includes 16 tourist resorts in beautiful environment, each providing supreme complete entertainment facilities, luxury water bath and breath-taking location.

Kempinski Hotel remoulded the concept of courtesy to ensure the courtesy team pay attention to each customer to the largest extent, not only meeting and exceeding customers' requirement but also wining recommendation on customer side. In order to highlight the importance of customers, the hotel has developed a series of courtesy projects, from business courtesy, private courtesy to conference courtesy, etc. Over the years, Kempinski has grown into an innovative and respectful luxury hotel group famous for providing the high standard service for international tourists in the world. Every hotel offers exquisite service and adds the specialties of the hotel to the local style.

New Hotel 新酒店

酒店概况

三亚海棠湾凯宾斯基酒店位于三亚海棠湾旅游度假区一号地，是规划中未来的世界顶级酒店的聚集区。本项目用地位于沙坝酒店区的最南端，西面是规划中的滨海景观大道，道路对面是大小龙江塘，东面紧临风景迷人的海棠湾，和蜈支洲岛隔海相望，南面是规划中的海洋公园和蜈支洲岛码头，地块交通景观优势明显。

Overview

Kempinski Hotel Haitang Bay Sanya is located in the 1st area of core of Sanya Haitang Bay Internatioanl Leisure and Vacation Zone—a cluster area of world-class hotel. The land is situated at the south end of Shaba Hotel Area, with planning Coastal Landscape Road on the west, Great and Lesser Longjiang Tang on the opposite side, charming Haitang Bay on the east looking over Wuzhizhou Island with the sea in between, planning marine park and Wuzhizhou dock on the south. The project enjoys obvious advantages of transportation.

New Hotel 新酒店

酒店特色

三亚海棠湾凯宾斯基酒店位于国家海岸——"海棠湾",拥有这条 20 km 的黄金海岸线上最原生态的沙滩。酒店内富有当地特色的黎苗文化元素与欧洲生活艺术完美结合。杜月笙的木艺术馆中收藏了大量珍贵的黄花梨古家具以及雕工精美的建筑遗迹,为来自全世界各地的宾客展现中国灿烂且悠久的木艺术文化。

Feature

Kempinski Hotel Haitang Bay Sanya is located on national Beach-Haitang Bay, embracing the most original beach on the 20 km golden coast line. Inside the hotel, there decorates characteristic local Li and Miao Cultural elements, in combination with European art of living. Du Yuesheng's Wood Art Gallary collects large number of precious yellow pear ancient furniture and architecture relics of exquisite carving techniques, demonstrating the spenlendid and long-lasting art and culture of China.

New Hotel 新酒店

酒店室内

酒店共有 396 间客房，包含 12 套水疗别墅房、1 栋 2 524 m² 的总统别墅。所有客房的设计都是为了追求更大的空间、更多的自然光线和更美的热带度假风情。从 55 m² 的豪华客房到 168 m² 的家庭套房均有单独的超大阳台，并在阳台上设有观景浴缸。酒店大堂穹顶由知名木雕大师手工雕刻而成，并加以箔金处理。设计师希望能有均匀的灯光洗亮天穹，在营造天光效果的同时，大堂天穹 LED 灯光控制能随着季节及早晚变化缓缓改变色温。酒店内收藏了诸多艺术品，其中不乏历经百年的古木雕、明清桌椅、建筑斗拱等。

Interior

The hotel has 396 guest rooms, including 12 spa villas, one presidential villa of 2,524 m², all are designed for larger space, more natural light and prettier tropical vacational atmosphere. From 55 m² luxury guest room to 168 m² family suite, all the rooms are equipped with super large terrace with landscape bath tub. The dome of the hotel lobby was hand carved by famous wood-carving master and treated with foil gold. Designers try to achieve even lights to light up the dome reminiscent of the sky while the LED control could adjust color and temperature according to the change of seasons and time of the day. The hotel collects many works of art, including wood carvings over 100 years, tables and seats of the Ming and Qing Dynasty and building brackets, etc.

酒店配套

■ 餐厅酒吧

通透宽敞的聚合吧惬意而悠闲，提供各式茶饮、咖啡、果汁、酒水及点心。客人也可在此畅饮酒店特调鸡尾酒或各式精选威士忌，伴随着现场表演度过一个美妙的夜晚。悦轩中餐厅设计优雅、现代，给客人营造轻松、愉快的就餐环境。临河而坐，观赏河景，或选择雅座包厢，品尝一系列新鲜、美味的生猛海鲜、粤式菜肴及海南特色美食。

■ 木艺术馆

由20世纪上海滩传奇人物杜月笙旧居改建的木艺术馆，古色古香，并由海水河道及18洞高尔夫推杆练习果岭环绕着，纯上海式的风格建筑宛如一个隐逸、隔世的天堂。明清时代的古老家具将在此向客人展出。

■ 国玺厅

国玺多功能厅位于酒店主楼中心，临河瞰海，是VIP会议，私人派对，社交场合及庆祝活动的理想选择。白墙黛瓦，古朴厚重，为典型的秦汉风格，建筑造型上方下圆，体现中国人"智欲圆而行欲方"的人生智慧。

Services and Amenities

■ Restaurants and Bars

The transparent and spacious Juhe Bar provides various drinks, coffee, juice, wine and snacks, giving the guests a pleasant and leisurely feeling. Guests could spend a wonderful night with tasting the specially made cocktail or Whisky and enjoying the live show. The design of Yuexuan Chinese Restaurant is elegant and modern, providing comfortable and pleasant dining environment for guests. Watching over the river, enjoying the beautiful riverside landscape or in the parlour box, guests could taste various delicious seafood dishes, Cantonese dishes and Hainan local cuisine.

■ Wood Art Gallary

The Wood Art Gallary was reconstructed from Shanghai's celebrity Du Yuesheng in the 20th century. It gives out ancient flavor, circled by waterways and golf putting green of 18 holes. The pure Shanghai style architecture seems like a far separated heaven, which displays the old furnitures of Ming and Qing Dynasty.

■ National Seal Hall

The multi-functional hall sits in the center of the main building, looking over the sea, which is an ideal choice for VIP meeting, private party, social events and celebrations. The black tiles – simple and decorous belonging Chinese to the classic Qin and Han style. The architectural layout is square up and round down, reflecting the life norm of "wisdom should be round while behavior should be square".

New Hotel 新酒店

The Naka Island, Phuket
— A Luxury Collection Resort and Spa

普吉岛纳卡岛度假酒店——豪华精选度假胜地

Keywords 关键词
Top Luxury 顶级奢华
Exotic Experience 异域情调
Charming Sea View 迷人海景

酒店地址：泰国普吉岛 32 Moo 5, Tambol Paklok, Amphur Thalang，纳卡娅岛
电　　话：(66)(76) 371 400

Address: 32 Moo 5, Tambol Paklok, Amphur Thalang, Naka Yai Island Phuket, Thailand
Tel: (66)(76) 371 400

品牌链接

作为喜达屋酒店与度假村国际集团旗下酒店，豪华精选（Luxury Collection®）是一系列可提供独特地道体验，为旅客留下珍贵难忘回忆的精选酒店。豪华精选致力于为各地旅游爱好者们打开一扇通往全球最激动人心和最令人向往的目的地的大门。每家酒店及度假村均独具特色、风情各异，堪称旅行目的地门户的豪华精选酒店将使旅客尽情领略原汁原味的当地文化和无限魅力。精美的装饰、豪华的布置、无与伦比的服务和先进的现代化便利设施将为旅客打造独特而丰富的入住体验。

About Luxury Collection®

As a subsidiary of Starwood Hotels, The Luxury Collection® is a selection of hotels and resorts offering unique, authentic experiences that evoke lasting, treasured memories. For the global explorer, The Luxury Collection offers a gateway to the world's most exciting and desirable destinations. Each hotel and resort is a distinct and cherished expression of its location; a portal to the destination's indigenous charms and treasures. Magnificent décor, spectacular settings, impeccable service and the latest modern conveniences combine to provide a uniquely enriching experience.

Special 专题

发展历程

豪华精选起源于 Compagnia Italiana Grandi Alberghi(CIGA) 公司，CIGA 成立于 1906 年，旗下拥有部分意大利最知名的豪华酒店，其中包括威尼斯的丹尼尔酒店。1985 年，Karim Aga Khan 王子取得了 CIGA 的控股权，并开始在意大利以外地区增扩豪华酒店的数量，特别是在西班牙和奥地利开展了大规模的收购活动。后来，CIGA 逐渐扩大豪华酒店所分布的地域范围，重点将撒丁岛上一些富有传奇色彩的酒店收归麾下，如狐狸湾大饭店、皮特利萨酒店和罗马契诺酒店。

1994 年，喜达屋酒店及度假村国际集团收购了 CIGA，并创办了豪华精选品牌。时至今日，豪华精选已在全球 30 个国家 / 地区设有 75 家世界一流的酒店及度假村，璀璨夺目。每家酒店的所在地都有其独到的引人之处，因此酒店的历史、建筑、艺术、家具和设施无不体现着本地深厚的文化底蕴，令人瞩目。豪华精选将当地习俗和特色与旅客体验完美地融合在一起，通过卓越的服务满足客人的每一个愿望。

History & Development

The Luxury Collection traces its heritage to Compagnia Italiana Grandi Alberghi (CIGA), founded in 1906, owner of some of Italy's most renowned luxury hotels, including the Hotel Danieli in Venice. In 1985, Prince Karim Aga Khan acquired a controlling interest in CIGA and began to expand the company's luxury hotel portfolio beyond Italy, most notably with purchases in Spain and Austria. CIGA later expanded in geographic focus to include legendary properties in Sardinia, including Hotel Cala di Volpe, Hotel Pitrizza and Hotel Romazzino. In 1994, Starwood Hotels & Resorts Worldwide, Inc. acquired CIGA and created The Luxury Collection designed to be the world's most renowned assemblage of hotels and resorts. Today, The Luxury Collection is a glittering ensemble of more than 75 of the world's finest hotels and resorts in more than 30 countries. All are noteworthy for their history, architecture, art, furnishings and amenities, highlighting the rich culture of each property's fabulous surroundings. The Luxury Collection seamlessly integrates local customs and location into the guest experience, while fulfilling every desire with extraordinary service.

酒店概况

普吉岛纳卡岛度假酒店位于 Andaman 海独具异国情调的小岛上。该度假酒店坐落在一个热带景观区，并提供带有私人泳池的豪华别墅，设有一个私人海滩、一个健身房和一个世界级的 Spa 水疗中心，客人可以享受免费的往返快艇接送服务。酒店距离普吉国际机场 25 分钟车程，距离珍珠姐妹纪念碑（Heroines Monument）20 km，距离普吉镇 25 km，距离 Patong 海滩 45 km。

Overview

Nestled among stunning beaches, lush coconut groves, with never-ending views of the emerald-green Andaman Sea and idyllic landscapes of the Phuket coastline, The Naka Island is an exclusive boutique resort on Naka Yai Island, located just off the Phuket coast. While just 25 minutes from Phuket International Airport, The Naka Island is accessible only by private speedboat, making it a uniquely private and intimate retreat.

Special 专题

酒店特色

酒店靠近宁静迷人的攀牙湾，很乐意为宾客提供多种水上运动，包括浮潜、帆板、帆船和皮划艇等。此外，宾客也可选择通过骑车或徒步游览这座美丽的海岛，或是静坐在洁白无瑕的海滩上欣赏壮阔迷人的海景。

Feature

Situated near the serene waters of Phang Nga Bay, we are pleased to offer a myriad of watersports, including snorkeling, windsurfing, sailing, and kayaking. Alternatively, explore the island's majestic landscapes with a cycling or hiking adventure, or simply relax on our spectacular ivory sand beach, marveling at the magnificent sea views.

酒店配套

■ 度假酒店泳池

在宽阔的淡水泳池中畅游的同时，宾客可尽情欣赏迷人的海景与攀牙湾美景。此外，相邻的热带池塘景色优美，与淡水泳池相映成趣，形成一条通往海滩的视觉走廊。露台上置放着众多休闲躺椅，宾客可舒躺在暮光之中，尽享惬意；或前往数步之外的海滩上参加振奋人心的休闲活动。

Services and Amenities

■ Swimming Pool

The large freshwater pool offers the singular experience of swimming while gazing out over the sparkling ocean and Phang Nga Bay. Meanwhile, an adjacent picturesque tropical pond enhances the pool area and creates an enchanting terraced visual leading to the beach. Peppered around the deck, chaise lounges invite guests to soak up the island sun while the beach offers an endless array of activities, just steps away.

■ 海滩

风景优美的海滩区配有太阳伞与太阳椅,并提供一切所需服务,其中包括食品与饮料,让宾客无忧无虑地沐浴阳光、欢度假期！宾客可通过酒店礼宾部安排非机动化的海滨活动,或选择在天然白沙滩上漫步徜徉,欣赏碧波荡漾的攀牙湾水景,感受宁静安逸的闲暇时光。

■ Beach

Furnished with umbrellas and sun loungers, the exquisite beach area provides everything necessary for a matchless day in the sun, including food and beverage service. Invigorating non-motorized seaside activities can be arranged through the Concierge, or guests can opt for a serene day on pristine white sands bordered by the emerald-green waters of Phang Nga Bay.

Special 专题

■ **美食**

酒店设有两家餐厅与一家酒吧,均位于海滩上,可观赏到攀牙湾、邻近岛屿及石灰岩悬崖等美景,带给宾客终生难忘的用餐体验。Tonsai 全日制餐厅供应自助早餐及美味无比的零点午餐与晚餐,随时满足宾客对美食的需求。此外,浪漫的 My Grill 餐厅以泰国南部的美食为灵感源泉,烹调出各色香烤创意菜肴,更有池畔低位用餐区打造超凡的用餐体验。位于酒店另一侧的 Z 酒吧则为宾客带来健康小吃和清爽日落鸡尾酒。

除此之外,宾客可在别墅内享受私密的海滩晚餐或有趣的烧烤活动。更有 24 小时客房内用餐服务保证宾客随时享用到美味佳肴,如传统西餐或当地特色美食等。应有尽有的饮品、美味可口的甜品与新鲜的当地水果让最挑剔的美食家也赞不绝口。独立的儿童菜单提供丰富的菜式,让小宾客们眼花缭乱。

■ Cuisine

Offering unforgettable dining experiences, the hotel's two restaurants and bar are situated beachside with ocean views of Phang Nga Bay, the surrounding islands, and dramatic limestone cliffs. Any time of day, Tonsai serves a buffet breakfast as well as an exceptional á la carte lunch and dinner. Meanwhile, the romantic My Grill enchants guests with its creative dinner menu of grilled options with a Southern Thai influence and charming sunken dining area surrounded by pools. For a healthy refreshment or captivating sunset cocktail, Z Bar is located on the opposite side of the resort.

Additionally, guests may request a private beachside dinner or BBQ in their villa, or sample our 24-hour in-room dining, which keeps them satiated with an array of gourmet offerings—including traditional Western dishes and local specialties for all three meals. An extensive beverage menu provides the perfect complement while desserts, both international favorites and local fruits, polish off the epicurean experience. A separate kids' menu means even our youngest guests have plenty to choose from.

■ 纳卡水疗中心

纳卡水疗中心坐落在迷人海岛上，波澜不兴的池塘随处可见，葱郁的花园美景令人难忘，堪比世外桃源。该水疗中心致力于为宾客带来焕发活力、激发内在潜能的健康体验。纳卡水疗中心毗邻健身中心，提供一系列针对头部与身体的特色植物护理，均由专业的理疗师提供。护理过程中所使用的产品均为来自意大利的舒适地带产品或纯正的泰国产品。宾客可选择在单人护理室中独享周到的服务，或在双人护理室与好友分享惬意的体验。此外，纳卡岛的酒店住客可免费使用一般的水疗设施。

■ Spa Naka

A singular island retreat dotted with placid ponds and boasting lush garden views, Spa Naka renews guests' strength and inner sense of well-being. Located near the Fitness Center, it offers an array of signature indigenous botanical treatments for the face and body, each performed by our expert therapists using exclusive Comfort Zone products from Italy or local Thai products. Select between one of the five single treatment rooms or share the experience in a double treatment room. Additionally, all in-house guests of The Naka Island have complimentary access to the general spa facilities.

■ 健身中心

健身中心俯览优美海景，分为伸展区和瑜伽区两部分，并设有泰诺健设备，其中包括班霸牌台阶器、椭圆机、健身车、蹦床及大量的重量训练器械与自由重量器械等。此外，中心还设有室外多功能凉亭，可用于进行瑜伽、普拉提与健身课程等。

■ Fitness Center

Overlooking the ocean, the Fitness Center includes a stretching and yoga area and is equipped with Technogym equipment, including a StairMaster, elliptical machine, stationary bike, trampoline, array of weight machines, and free weights. Additionally, there is an outdoor multi-purpose sala that can be used for yoga, pilates, and fitness classes.

| Special 专题

Jumeirah Vittaveli
卓美亚维塔维丽度假酒店

Keywords 关键词
- Modern Luxury 现代奢华
- Innovative 创新
- Water Gym 水上运动
- Talise 泰丽丝

品牌链接

卓美亚酒店及度假村是世界上最豪华最具创新性的项目，至今已经赢得了众多的国际旅游奖。集团成立于1997年，旨在通过建立一个世界一流的豪华酒店和度假村组合成为酒店行业的领头羊。在此成功的基础上，2004年卓美亚集团成为迪拜的一员——在迪拜开展一系列商业活动和项目，符合集团新的成长和发展要求。

About Jumeirah

Jumeirah Hotels & Resorts are regarded as among the most luxurious and innovative in the world and have won numerous international travel and tourism awards. The company was founded in 1997 with the aim to become a hospitality industry leader through establishing a world class portfolio of luxury hotels and resorts. Building on this success, in 2004 Jumeirah Group became a member of Dubai Holding – a collection of leading Dubai-based businesses and projects – in line with a new phase of growth and development for the Group.

酒店地址：马尔代夫南马累环礁流星岛
电　　话：+960 665 8111

Address: Bolifushi Island, South Male Atoll, Republic of Maldives
Tel: +960 665 8111

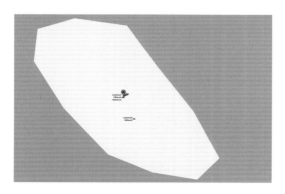

Special 专题

酒店管理

卓美亚维塔维丽度假酒店关注气候，注重环保，努力保护酒店周围易被破坏的生态系统。通过设立学徒奖、在酒店拍卖儿童艺术品和卓美亚卓越奖（旨在表彰杰出的学术成就）等举措，为儿童教育和培训中心(ETCC)以及Maafushi学校提供资金和非资金支持。

为保护自然景观，酒店会定时清洁周围的群礁和沙滩，并已使用有氧污水处理厂，避免污水流入海洋。此外，还成立了环境委员会，继续寻找新的方法，解决这一宝贵栖息地独特的环境和社会需求。

Management

At Jumeirah Vittaveli it is passionate advocate of climate awareness, and work tirelessly to support the community and protect the delicate ecosystem that surrounds the island hideaway. It offers financial and non-financial support to Education and Training Centre for Children (ETCC) and the Maafushi School through initiatives like an apprenticeship award, a children's art sale at the hotel, and the Jumeirah Award of Excellence, which recognises outstanding academic achievement.

To protect the natural landscape, it cleans the reefs and beaches around the hotel, and has invested in the use of an aerobic sewage plant, which returns no effluent to the ocean. Finally, it has established an environmental committee to continue to find new ways to address the unique environmental and social needs of the precious habitat.

酒店特色

卓美亚维塔维丽度假酒店位于马尔代夫仅有的 200 个有人居住的岛屿之一,集便利交通、奢华服务和私密性于一体,为顾客打造独一无二的放松体验。从首都马累的马累国际机场,只需短短的 20 分钟,便可乘船抵达酒店。还有提供高效的机场往返接送服务。

Feature

One of only 200 populated islands in the Maldives, Jumeirah Vittaveli is a resort that combines accessibility with luxury and privacy. From the capital of Male, you can reach in just a short 20-minute boat ride from Male International Airport. It will be sure your transfer is handled seamlessly.

Special 专题

酒店室内

酒店共有44间海滩别墅和套房、46间礁湖别墅和套房，三间餐厅和一间海滩鸡尾酒吧。每栋别墅均配有私人泳池，拥有丰富的水上运动，包括深海捕鱼、潜水、皮划艇和风筝冲浪；泰丽丝水疗中心，设有岛上和水上护理室。

Interior

The hotel has 44 beach villas and suites, 46 lagoon villas and suites, three restaurants and a beachside cocktail bar, private swimming pool in every villa, full range of water sports including deep sea fishing, diving kayaking and kite surfing and Talise spa with treatment rooms on land and over water.

Special 专题

酒店配套

■ 潜水和水上运动

卓美亚维塔维丽度假酒店拥有各种独特的水下奇观，绝对是宾客逗留期间不容错过的享乐胜地。岛屿四周水域环绕，从美丽的珊瑚礁到惊险刺激的潜水区，各种休闲方式一应俱全，带宾客踏上精彩纷呈的冒险之旅。酒店的"Best Dives"潜水和水上运动中心获得了专业潜水教练协会 (PADI) 认证，提供多种多样的课程及探险训练。宾客可以乘坐皮划艇巡游风平浪静的海面，也可寻找欣赏美丽珊瑚群的最佳地点，或是在夕阳西下之际享受垂钓之乐。

Services and Amenities

■ Diving & Water Sports

Jumeirah Vittaveli is home to unique underwater wonders which are an absolute must-see during your stay. The waters that surround the island offer incredible adventure, from beautiful coral reefs to inspirational dive sites.

PADI certified Best Dives dive and water sports centre provide a wide array of courses and expeditions. Speak to the professionals on your preferences and they will help select an activity that's right for you – be it cruising down the calm waters on a kayak, finding the best spots to discover the coral gardens, or enjoying a sunset fishing trip.

■ 泰丽丝水疗中心

Vittaveli 有许多美好的寓意——广阔的空间、永恒的光芒。这些完全符合这座岛屿的特点。Vittaveli 泰丽丝水疗中心的护理秉承这些光与空间元素,让客人享受一片纯净并产生共鸣。护理所用的材料在当地通过传统方式获得,确保万物中生命愈合的能量得到尊重,并通过特有的产品和护理传递给客人。

专业的水疗师提供特色护理,为顾客带来全面的身心呵护,使顾客重归自然。专为情侣打造的浪漫空间,使顾客放松心情,尽享私密时刻。种类丰富的水疗项目亦可根据需求为顾客量身打造。

■ Talise Spa

The word Vittaveli has many beautiful meanings – vastness of space and eternal light. These meanings fit perfectly with the island. Treatments at Talise Vittaveli are inspired by these same elements of light and space connecting with the purity and vibration of our ingredients. These ingredients are hand-harvested by local communities using traditional farming methods to ensure healing life energy held in all living things will be honoured and handed down to you through our unique products and treatments.

A holistic approach to wellbeing reconnects you with nature, with signature treatments delivered by the expert spa therapists. Romantic spaces for couples offer intimate moments of true relaxation, while our extensive spa menu offers individual treatments tailored to your needs.

| Special 专题

■ 健身

高科技水上健身房可为顾客提供各种挑战运动,让顾客在入住期间达到身心平衡的健康状态。一边使用高科技设备进行锻炼,一边欣赏印度洋秀美迷人的全景,畅享健康愉悦的健身体验。在设施完备的健身中心和24小时健身中心参加免费私人训练课程,为顾客的健身计划注入新的活力。此外,海滩训练选择、循环训练和伸展课程还将帮助顾客以最适合自己的方式保持充沛活力。想要恢复精力的人士则可选择私人和团体瑜珈课程。

■ Fitness Center

The state of the art over-water gym provides a challenging and rewarding space to enable a healthy balance during your stay. Draw inspiration from panoramic views of the Indian Ocean as you use first class equipment designed to help you achieve your fitness goals. Breathe new life into your fitness regime with a complimentary personal training session in the well-appointed health club and 24-hour fitness centre. There's also beachside training options, circuit training and stretch classes to help you stay active in the way that suits you. For those interested in restoring the balance of body and mind, private and group yoga classes are available. The adjacent 24-hour Fitness Center is equipped with state-of-the-art gym equipment, and a stunning 25 metre (82 foot) horizon-edge lap pool overlooks the city with a windowed poolside relaxation terrace.

■ **婚礼与蜜月**

对于人生中最重要的婚礼而言,马尔代夫清澈的海水、细软洁白的沙滩和湛蓝天空令人心动。岛上的自然风光为婚礼营造出如诗如画的优美环境,而酒店的服务和设施更可为顾客带来无与伦比的奢华体验。

经验丰富的婚礼协调人员和活动策划人员将为顾客打点好一切。酒店人员将以饱满的热情提供最贴心最专业的服务,让顾客用心感受每一个珍贵时刻。顾客只需尽情享受如天堂般美丽的环境,全身心投入人生中最特别的时刻。

■ Wedding and Honeymoon

When it comes to special occasions, the clear waters, white sandy beaches and blue space of the Maldives hideaway are hard to beat. The island's natural beauty is an idyllic backdrop to any special occasion while the hotel's services and amenities add a true touch of Jumeirah luxury.

The experienced team of wedding coordinators and event planners will take care of everything for you. They will offer support, expertise and enthusiasm, leaving you free to savour each special moment. All you need to do is soak up the paradise around you and celebrate one of life's most special moments.

Jumeirah Dhevanafushi
卓美亚德瓦纳芙希度假酒店

Keywords 关键词

Modern Luxury 现代奢华

Art Design 艺术设计

Water Gym 水上健身

Superior and Innovative 卓越创新

品牌链接

卓美亚酒店及度假村是世界上最豪华最具创新性的项目，至今已经赢得了众多的国际旅游奖。集团成立于1997年，旨在通过建立一个世界一流的豪华酒店和度假村组合成为酒店行业的领头羊。在此成功的基础上，2004年卓美亚集团成为迪拜的一员——在迪拜开展一系列商业活动和项目，符合集团新的成长和发展要求。

About Jumeirah

Jumeirah Hotels & Resorts are regarded as among the most luxurious and innovative in the world and have won numerous international travel and tourism awards. The company was founded in 1997 with the aim to become a hospitality industry leader through establishing a world class portfolio of luxury hotels and resorts. Building on this success, in 2004 Jumeirah Group became a member of Dubai Holding – a collection of leading Dubai-based businesses and projects – in line with a new phase of growth and development for the Group.

酒店地址：马尔代夫卡夫阿里夫环礁梅拉德岛
电　话：+960 682 8800

Address: Meradhoo Island, Gaafu Alifu Atoll, Republic of Maldives
Tel: +960 682 8800

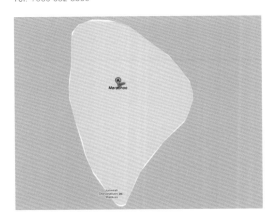

酒店管理

酒店的成功和可持续发展遵循以下几个原则：客户至上，努力提高不断超越他们的期望；将酒店的经典案例作为日常活动的指导原则和核心标准；学习和执行酒店发展积累起来的一切优秀文化；提供始终如一的卓越和创新的产品和服务。

Management

The success and sustainability of the hotel is achieved by: Making customers the first priority and striving constantly to exceed their expectations; Applying the hotel Hallmarks, Guiding Principles and Core Standards in the day-to-day activities; Leading by example and role-modeling a culture of excellence in everything they do; Providing consistently superior and innovative products and services.

酒店概况

卓美亚德瓦纳芙希度假酒店位于印度洋中心的马尔代夫南端，是体验宁静、愉悦生活的一方净土。这处迷人的避世圣地横越两个岛屿，让顾客远离喧嚣纷扰，品味珍贵时光。苍翠繁茂的植物、晶莹剔透的海水和绵延迂回的白色沙滩，营造出如诗如画的绝美环境。卓美亚德瓦纳芙希度假酒店堪称一片宁静天堂。

这座避世之岛四周萦绕着印度洋的异域之美，散发着奢华幽谧的气息。卓美亚德瓦纳芙希度假酒店距首都马累 400 km，客人可乘坐短途国内航班或海上飞机到达。

Overview

Nestled at the south end of the Maldives in the heart of the Indian Ocean, Jumeirah Dhevanafushi was created for life's tranquil pleasures. Stretched across two islands, its enchanting hideaway offers everything the client need to get away from it all. Lush foliage, crystal blue waters and stretches of white sandy beach create a sanctuary. It is no exaggeration to call this paradise.

Surrounded by the exotic beauty of the Indian Ocean, the island hideaway surely boasts some of the most remote luxury to be found on earth. Located 400 km from the capital Male, guests can arrive at Jumeirah Dhevanafushi by a short domestic flight or seaplane transfer.

| Special 专题

酒店特色

酒店提供各种舒适的设施和服务,为顾客打造与众不同的入住体验。顾客可在泰丽丝水疗中心的精致餐厅享用丰富美食,或仅仅在珊瑚礁中畅游一番,恢复活力。

Feature

The hotel offers an extensive list of comforts to make the clients' stay special. It stimulates the senses in one of its restaurants, at the Talise Spa, or simply reinvigorate with a swim among the coral reef.

酒店配套

■ 美食休闲

卓美亚德瓦纳芙希度假酒店的三家顶级餐厅提供琳琅满目的精致菜肴，打造愉悦感官的奢华体验。所有菜品均以最新鲜的本地食材精心烹制而成，色香味俱全，令人难以取舍。菜肴的烹制方法十分简单，突出原汁原味的口感，各种令人垂涎的美味结合如画的岛屿风景，带来美妙难忘的体验。

从新鲜可口的美食到热情周到的待客之礼，卓美亚德瓦纳芙希度假酒店将使顾客的每次用餐体验都独一无二。

■ 庆典及会议

马尔代夫可提供各种婚礼与蜜月场所。但只有一个场所可以为顾客提供激动人心的奢华和卓越非凡的服务，并竭尽所能地满足他们的需求和愿望。

那便是卓美亚德瓦纳芙希度假酒店，它将为顾客爱情故事中的永恒时刻提供美丽非凡的陪衬环境。无论顾客在计划婚礼还是蜜月，酒店都将悉心关注每一个细节。酒店将以饱满的热情提供最贴心、最专业的服务，让顾客用心感受每一个珍贵时刻。

酒店是专为浪漫而造的海上世外桃源。这里有洁白细软的沙滩、葱郁嫩绿的草木和水晶般透明的湛蓝海水，顾客的童话梦想将在此成为现实。

Services and Amenities

■ Dining and Entertainment

Delight your senses with the sumptuous menus of three restaurants at Jumeirah Dhevanafushi. Every menu offers a tough choice, dishes incorporating the freshest locally sourced ingredients available. The approach to dining is simple; the hotel let flavour speak for itself, creating combinations as memorable as the island setting.

From the succulence of the food to its welcoming hospitality, people wll find every meal exceptional at Jumeirah Dhevanafushi.

■ Celebrations and Other Events

When choosing a wedding or honeymoon venue in the Maldives, the choices are infinite. But only one venue can offer a truly seamless blend of inspirational luxury, exemplary service and total focus on your needs and wishes.

Jumeirah Dhevanahushi is an exceptionally beautiful setting for this momentous scene in your love story. Whether the clients are planning their wedding, honeymoon or both, it will be there to take care of every detail. It will offer support, expertise and enthusiasm, leaving the client free to savour each special moment.

The island hideaway exists for romance. Bring your vision to life with the white sandy beaches, lush greenery, azure crystal waters and all of the facilities of its elegant resort.

■ 健身

高科技水上健身房可为顾客提供各种挑战运动，让顾客在入住期间达到身心平衡的健康状态。一边使用高科技设备进行锻炼，一边欣赏印度洋秀美迷人的全景，畅享健康愉悦的健身体验。在设施完备的健身中心和 24 小时健身中心参加免费私人训练课程，为顾客的健身计划注入新的活力。此外，海滩训练选择、循环训练和伸展课程还将帮助顾客以最适合自己的方式保持充沛活力。想要恢复精力的人士则可选择私人和团体瑜珈课程。

■ Fitness Center

The state of the art over-water gym provides a challenging and rewarding space to enable a healthy balance during the clients' stay. Draw inspiration from panoramic views of the Indian Ocean as the client use first class equipment designed to help he/she achieve his/her fitness goals.

■ 水疗中心

泰丽丝水疗中心将其奢华本色融入到了卓美亚德瓦纳芙希度假酒店的静谧世界之中,利用马尔代夫海岸线的绝美景观为其特色水疗体验注入新能量。专业的水疗师将以特色护理带给顾客身心和谐的美妙体验。

水疗不再只是愉快的体验,而是发现自我的全新旅程。每项护理都根据个性化需求特别设计,带来的远不止是身心的放松。泰丽丝将为顾客开启极致放松之旅,带顾客遨游内心宁静和谐的港湾。专为情侣打造的浪漫空间可让顾客彻底放松,尽享温馨私密时刻。丰富多样的水疗项目更可根据顾客的需求量身打造。

■ Talise Spa

The luxury of Talise Spa finds a particularly tranquil setting at Jumeirah Dhevanafushi, where its signature spa experience overlooks the Maldivian coastline. A holistic approach to wellbeing reconnects people with nature, with signature treatments delivered by its expert spa therapists.

The experience becomes a journey of self discovery rather than just a feel-good experience. Each treatment is personalised to individual requirements and gives people more than just a feeling of relaxation. Talise unfolds a sense of deep tranquility and balance for complete holistic rejuvenation. Romantic spaces for couples offer intimate moments of true relaxation, while our extensive spa menu offers individual treatments tailored to your needs.Whether it's a deep tissue massage or a purifying facial, we'll take you on a journey to meet a calmer, more relaxed you.

Special 专题

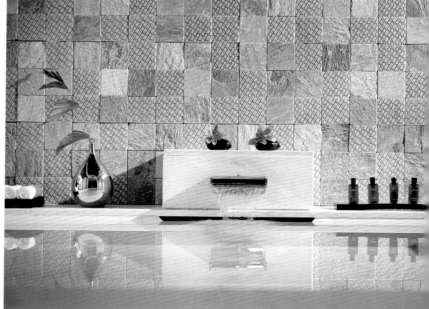

Sheraton Sanya Haitang Bay Resort

三亚海棠湾喜来登度假酒店

Keywords 关键词

Coastal Resort 滨海度假

Tropical Garden 热带花园

Multiple Amenities 多元配套

Elegant Seascapes 雅致海景

酒店地址：中国海南三亚海棠湾海棠北路
电　话：(86)(898) 3885 1111

Address: Haitang North Road, ,Haitang Bay, Sanya, Hainan, China
Tel: (86)(898) 3885 1111

品牌链接

喜来登酒店与度假村集团（Sheraton Hotels and Resorts）是喜达屋（Starwood）酒店集团中最大的连锁旅馆品牌，而它也是集团中第二老的酒店品牌（最老牌的是威斯汀）。今日的喜来登品牌是在1937年出现的，当时两位企业家Ernest Henderson以及Robert Moore在马萨诸塞州的斯普林菲尔德成立了第一家喜来登酒店。1945年喜来登成为第一家在纽约证券交易所挂牌上市的连锁酒店集团。1995年福朋喜来登品牌成立，喜来登希望以合理的价格提供全方位的服务；当时很多规模较小的喜来登酒店都被改名为福朋喜来登。1998年喜达屋集团以高于希尔顿的出价收购了喜来登品牌。在喜达屋的管理领导下，喜来登开始创建更多的酒店以扩大其品牌影响力。

About Sheraton

Sheraton Hotels and Resorts is the largest hotel chain brand of the Starwood Hotels and Resorts Worldwide Inc. and second oldest brand (Westin being the oldest).The origins of the sheraton dated back to 1937 when Ernest Henderson and Robert Moore acquired their first hotel in Springfield, Massachusetts. In 1945, it was the first hotel chain to be listed on the New York Stock Exchange. In 1995, Sheraton introduced a new, mid-scale hotel brand Four Points instead of Sheraton Hotels, Which provides a comprehensive service at a reasonable price. In 1998, Starwood Hotels & Resorts Worldwide, Inc. acquired Sheraton, outbidding Hilton. Under Starwood's leadership, Sheraton has begun renovating many existing hotels and expanding the brand's footprint.

| Special 专题

酒店概述

三亚海棠湾喜来登度假酒店位于享有"国家海岸"美誉的海棠湾中心地带，与蜈支洲岛隔海相望，并毗邻诸多著名旅游景点及多个高尔夫球场。这里阳光灿烂，海水湛蓝，椰林婆娑，民风淳朴……是休闲度假及举办会议的理想之地。酒店拥有500间宽敞的客房与套房，面朝叹为观止的海景或葱郁园景。所有客房和公共区域配备有线及无线高速互联网连接。

Overview

Sheraton Sanya Haitang Bay Resort lies along the striking Haitang Bay labeled National Seashore, facing Wuzhizhou Island across the sea and adjacent to many famous tourist attractions and golf courses. It is an ideal place for vacation and conference, full of palms, coconuts, white beaches and abundant sunshine. The hotel hosts 500 spacious guest rooms and suites with excellent sea views or garden views. Broadband/wireless internet access is set in all guestrooms and public areas.

酒店特色

酒店拥有丰富多彩的休闲娱乐选择，包括各种趣味盎然的户外活动和海滩上的彻底放松。客户可以在这里体验一系列机动水上活动和水上运动，还可前往水疗中心享受无与伦比的个性化服务和放松护理，体验先进一流的健身中心和网球场。

专为社交目的而设的"随心连动@喜来登，体验在微软"可让客户与他人轻松交流，同时利用免费无线高速上网接入和联网电脑工作站收发电子邮件、搜索当地景点、甚至打印登机牌。酒店还设有商务中心，可帮助客户随时跟进各项工作。儿童俱乐部的每日综合计划可让孩子们在整个住宿期间都能尽享愉悦体验，其丰富多彩的活动从简单的绘画课程到脸部彩绘，一应俱全。入住三亚海棠湾喜来登度假酒店的客人还能享用其他酒店的服务设施，如屋顶画廊和雪茄吧，位于本酒店隔壁的姊妹酒店：三亚御海棠豪华精选度假酒店。

Feature

Plenty of recreational opportunities are offered here, including fun activities and total relaxation on the beach. The customers could enjoy a range of motorized water activities and water sports, at the spa with unparalleled personalized service and treatments, and get a thorough workout at the state-of-the-art fitness centre or on the tennis court.

The Link @ Sheraton Experienced with Microsoft® is designed to serve needs to balance work and play while also sharing the travel experience with each other. Designed as a social destination, the Link @ Sheraton invites customers to interact with each other while checking email, researching local attractions, and even printing boarding passes using free wireless High Speed Internet Access and Internet-enabled computer stations. A business centre is also available to help stay on top of work. Younger guests are invited to experience all the fun and games of the renowned Sheraton Kids Club. Offering a comprehensive daily program, the club keeps children entertained throughout their stay with activities ranging from simple drawing class to face painting. Guests of the Sheraton Sanya Haitang Bay Resort are allowed to access the facilities—such as the Top Gallery and the Cigar Bar of its sister property next door: The Royal Begonia, Sanya.

| Special 专题

酒店配套

■ 餐饮

三亚海棠湾喜来登度假酒店设有四间美食餐厅,为顾客供应各种风味佳肴,包括中式、亚洲、西式和欧式等,一应俱全。顾客可选择品尝本地的特色美食——包括可观海景的露天休闲小吃或品种丰盛的海鲜大餐。专业厨师以来自世界各地的丰富知识与工作经验打造令人难忘的餐饮体验。酒店的两间酒吧将为顾客奉上专业调制的鸡尾酒和上等葡萄酒。此外,还提供24小时客房送餐服务,包括由各餐厅提供的精选美食、丰盛佳酿和颇受欢迎的本地小吃及面点。

Services and Amenities

■ Dining

Sheraton Sanya Haitang Bay Resort is home to four restaurants offering a range of cuisines, including Chinese, Asian, Western, and continental, as well as numerous local delicacies— a casual al fresco snack overlooking the ocean or extensive selection of seafood. The specialty chefs bring with them a wealth of knowledge and experience gained from working around the world, ensuring that every meal is a memorable culinary experience. Also, two bars offer expertly mixed cocktails and fine wines. And 24-hour room service is available, offering selections from all of our restaurants, an ample beverage list, and popular snacks and local noodles.

Special 专题

■ 会议与活动

三亚海棠湾喜来登度假酒店拥有迷人的海滨美景、葱郁的热带花园和灵活的多功能空间，无疑是举办会议和活动的理想场所。酒店拥有总面积超过 3 600 m² 的 10 间会议室和面积达 1500 m² 且可分为三个同等大小区域的豪华宴会厅。此外，酒店还设有宽敞的迎宾空间和户外空间——如数个泳池区和海滨花园等，是举办鸡尾酒会和婚礼宴会的绝佳之地。所有会议场所均配有先进一流的高科技设施，包括无线高速上网接入、视频会议服务和液晶投影仪。接待服务台可提供技术协助和个性化支持。

■ Meetings & Events

Stunning beachfront views, lush tropical gardens, and flexible function space are just a few reasons to host events at the Sheraton Sanya Haitang Bay Resort. It offers more than 3,600 m across ten meeting rooms and the 1,500 m Grand Ballroom, which can be divided into three equal areas. Also, plenty of pre-function and outdoor space—such as pool areas and beachside gardens—is available for idyllic cocktail parties and wedding receptions. All meeting facilities feature the latest technology, including wireless High Speed Internet Access, video teleconferencing services, and LCD projectors. A hospitality desk provides technical assistance and personalized support.

Special 专题

Special 专题

■ 客房及套房

460 间宽敞客房及 40 间套房均秉承舒适至上的设计理念,面积从 50 至 150 m² 不等,可欣赏到壮观迷人的海洋美景和泳池风光。每间客房都配有完善的服务设施、静谧的现代化装潢和贴心周到的喜来登细节:独具特色的喜来登甜梦之床™铺有豪华床垫和舒适羽绒被,此外卧室内还设有一台液晶纯平电视和一张宽大的办公桌。客户可通过有线及无线高速上网接入与亲友和同事保持紧密联系。

客户可在舒适的花园景观客房悠然地欣赏园林美景。或是走出泳池景观客房或尊贵豪华海景客房的宽敞阳台,饱览壮观景致。尊贵泳池直通套房则设有专属的私人泳池。室外空间拥有宁静的氛围,是享受浪漫的理想之地。尊贵豪华海景套房是酒店面积最大的住宿空间,可在此欣赏壮观而难忘的海边日出。

■ Rooms and Suites

Boasting breathtaking ocean, pool views and ranging in size from 50 to 150 square metres, 460 spacious guest rooms and 40 suites have been designed with comfort in mind. Each one is equipped with ample amenities, serene modern décor, and familiar Sheraton touches, signature Sheraton Sweet Sleeper™ Bed with its plush mattress and cozy duvet, as well as an LCD Flat Screen Television and a generous work desk. Stay connected to friends, family, and colleagues with wired and wireless High Speed Internet Access.

People could admire a soothing perspective of the garden from the comfort of a Garden View Room. Or step out onto the large balcony of a Pool View Room or Grand Deluxe Sea View Room to take in the sights. The Grand Pool Access Suite includes an exclusive private swimming pool. The outdoor area's tranquil ambience provides a thoroughly romantic setting. The luxurious Grand Deluxe Sea View Suite is our largest accommodation. In this space, watching the spectacular sunrise across the sea is an unforgettable experience.

Special 专题

Sheraton Rhodes Resort
罗得岛喜来登度假酒店

Keywords 关键词

Aegean Sea 爱琴海

Field Experience 场所体验

Seaview Room 海景客房

Luxury Style 奢华格调

酒店地址：希腊南爱琴海罗得岛 Ialyssos 大道
电　话：(30)(22410) 75000

Address: Ialyssos Avenue, Rhodes, South Aegean 85100 Greece
Tel: (30)(22410) 75000

品牌链接

喜来登酒店与度假村集团（Sheraton Hotels and Resorts）是喜达屋（Starwood）酒店集团中最大的连锁旅馆品牌，而它也是集团中第二老的酒店品牌（最老牌的是威斯汀）。喜来登的酒店型态有许多种，从一般的商业旅馆到大型度假村都有；喜来登品牌一直力图维持高品质形象，在世界上的喜来登酒店有超过一半被当地机关评选为五星级酒店。喜来登酒店据点分布极广，遍布五大洲，从香港到斯里兰卡到埃及津巴布韦等国都可见其旅馆。喜来登总部在美国纽约的白原市。

喜达屋酒店都有良好的选址，主要分布在大城市和度假区。集团酒店选址的标准是：所在区域的发展史表明，该地区对提供全方位服务的豪华高档酒店有大量、持续增长的需求。作为酒店业豪华高档细分市场中最大的酒店集团，喜达屋酒店的规模有力地支持它的核心市场营销和预定系统。喜达屋酒店在把重点放在豪华高档细分市场同时，其各种品牌分别侧重于该市场中不同的二级市场。喜达屋酒店在赌场业也占据着重要的位置，它主要是通过 Caesars 品牌来经营此业务。喜达屋酒店为休闲度假旅游者提供着宾至如归（home-away-from-home）的服务。

About Sheraton Hotels and Resorts

Sheraton Hotels and Resorts is Starwood Hotels and Resorts Worldwide's largest and second oldest brand (Westin being the oldest). It occupies many types of hotel properties, ranging from general commercial hotels to large-scale resorts with high-quality image; more than half of Sheraton hotel worldwide are recognized as five-star hotels, with a widespread distribution across five continents i.e. Hong Kong, Sri Lanka, Zimbabwe etc. It headquarters in White Plains, New York.

Starwood hotels are located mainly in big cities or resort districts, in accordance with the site selection standards of group hotel that the selected districts own an increasing and great demand for full-serviced luxurious high-end hotels. As the greatest hotel group in the luxurious high-end market of hotel industry, Starwood holds its coral market promotion and booking system supported by its scale. It lays its emphasis on luxurious high-end market and its brands are targeted at the secondary markets. It also has a great achievement in casino industry, mainly represented by Caesars brand. Sheraton hotels address themselves to offer the visitors services of home-away-from-home.

Special 专题

酒店开发

喜来登酒店在收购罗德岛帝国酒店之后对其进行了翻新，改名罗得岛喜来登度假酒店，于 2010 年 5 月 1 日开业，该酒店是希腊的第一个喜来登酒店。

Development

The Sheraton Rhodes Resort, the first Sheraton hotel in Greece, was officially launched on May 1 2010 after the ownership of the former Imperial Rhodes Hotel was taken by Sheraton Hotels.

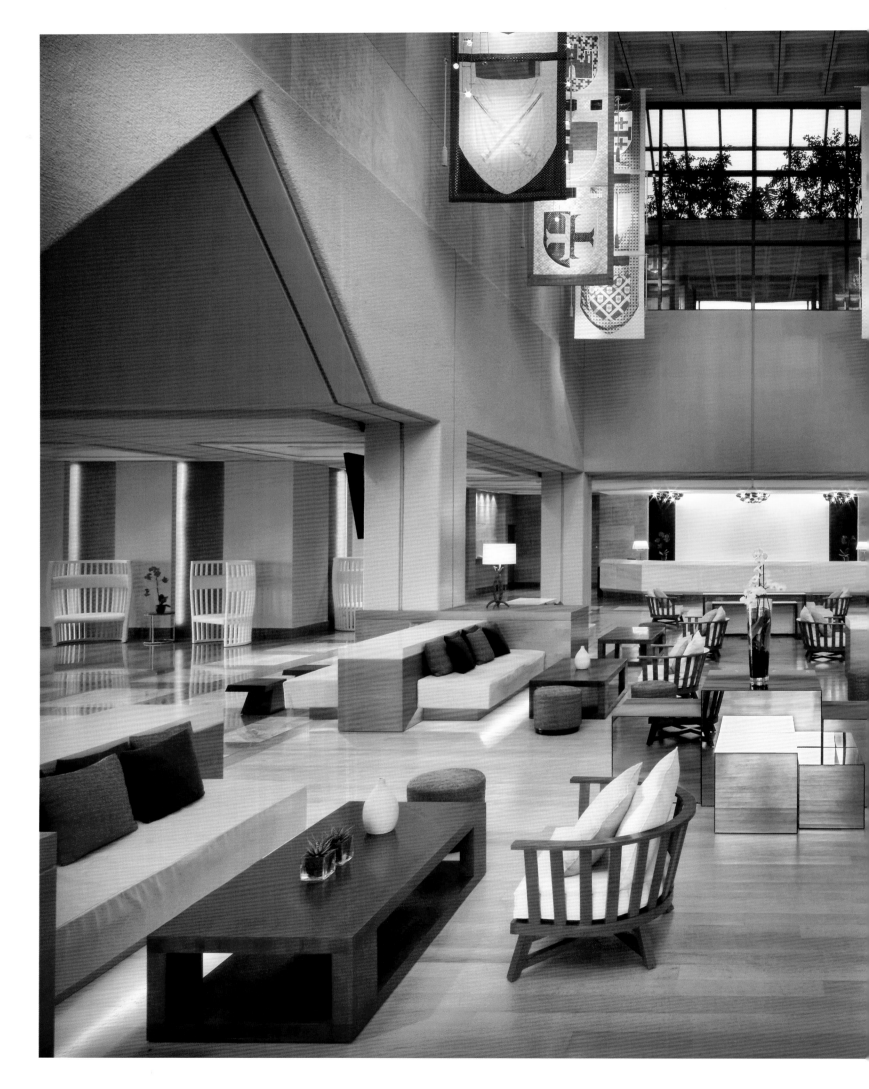

酒店概况

酒店坐落在罗德岛上,距罗德岛古城仅数公里之遥,距罗德岛国际机场仅 20 分钟车程。焕然一新的罗德岛喜来登度假酒店能够为来自全球各地的旅行者、家庭与蜜月游客提供最佳度假设施。

Overview

Located on the island of Rhodes, only a few kilometers from the historic town of Rhones and less than twenty minutes' drive from Rhodes International Airport, the refurbished Sheraton Rhodes Resort offers global travelers, families and honeymooners alike the best of resort amenities.

Special 专题

酒店特色

罗德岛喜来登度假酒店为客人提供五星级度假胜地所能提供的一切服务,让客人在入住期间,感受最优质的服务、最友好的氛围和最与众不同的舒适。

Feature

Sheraton Rhodes Resort provides all the services guests would expect from a five-star resort. During their stay the guests find top quality service and friendly atmosphere, as well as most distinguished comfort. High Speed Internet Access is available in guest rooms as well as in the public areas.

酒店配套

■ 餐饮

罗得岛喜来登度假酒店餐饮种类众多,从传统的希腊风味到国际口味,以供客人选择。酒店的行政主厨还为会议和奖励旅游设计了一些创意方案,如在酒店的泳池旁举行烧烤餐和希腊乡村风格的派对、甚至是在户外的卡利西亚泉水旁承办晚宴。酒店包括以下几个酒店和餐厅: L'Onda, Mediterraneo, Castellania, Lounge Bar 和 Poolbar。

Services and Amenities

■ Dining

An excellent choice of restaurants is available at the resort, the cuisine ranging from traditional Greek to international tastes. The Executive Chef has designed many creative options for conferences and incentives including a welcome barbeque and Greek Village party around one of the Hotels pools, or even an outdoor catering Gala Dinner at Kallithea Springs. Sheraton Rhodes Resort hosts the following Bars & Restaurants: L'Onda, Mediterraneo, Castellania, Lounge Bar and Poolbar.

■ 房间

罗得岛喜来登度假酒店有401间客房,享有山景、海景或园景,其中20间普通套房和12间豪华套房面积将近140 m²,尊享广阔蔚蓝的爱琴海海景。房间类别共八种:标准房、高级海景房、豪华客房、普通套房、爱琴海岸套房、家庭房、海景家庭房和家庭套房。所有的房间都配有甜梦之床。

■ 会议

罗得岛喜来登度假酒店提供一系列多功能空间,包括一个帝王宴会厅和14个大小不一的会议室,会议总面积达2 627 m²。酒店可承办同时容纳650名与会者的会议和800名来宾的宴会。所有会议设施自然采光,整个会议场所都有无线网络覆盖。此外还有全方位运作的商业中心和随心连动@喜来登,客人可以在这个网络与氛围俱佳的环境中查看邮件,无需额外付费。

■ Rooms

Sheraton Rhodes Resort offers 401 spacious guestrooms and suites with a variety of sea, mountain, or garden views. 20 Junior Suites and 12 Suites of up to 140 m², offer expansive views over the azure waters of the Aegean sea. The 401 rooms are divided into eight categories: Classic Room, Superior Sea View Room, Deluxe Room, Junior Suite, Aegean Suite, Family Room, Family Room with Sea View and Family Suite. All rooms feature the Sweet sleeper bed.

■ Meetings

The Sheraton Rhodes Resort has a selection of multi-purpose function spaces including an Imperial Ballroom and 14 meeting rooms which vary in size, offering a total Conferencing area of 2,627 square meters. The hotel is able to accommodate up to 650 delegates for a Conference and up to 800 guests for a banquet. The facilities offer natural daylight, wireless internet throughout our meeting spaces, a fully operational business center and "the Link at Sheraton" where guests of the hotel may check their e-mails in comfortable surroundings at no additional charge.

The Ritz Carlton Hotel, Okinawa | 冲绳丽思卡尔顿度假酒店

Keywords 关键词

Gusuku Ancient Architecture Style 冲绳古建风格

Golf Landscape 高尔夫景观

Refined and Peaceful 雅致静谧

酒店地址：日本冲绳岛
电　话：+81 980 43 5555

Address: Kise, Nago, Okinawa, Japan
Tel: +81 980 43 5555

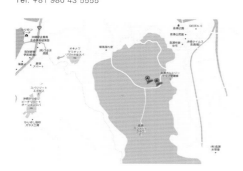

品牌链接

丽思卡尔顿酒店（Ritz-Carlton）是一个高级酒店及度假村品牌，现时拥有超过70个酒店物业，分布在24个国家的主要城市。丽思卡尔顿酒店由附属于万豪国际酒店集团的丽思卡尔顿酒店公司（Ritz-Carlton Hotel Company）管理，现雇用超过38 000名职员，总部设于美国马里兰州，靠近华盛顿特区。

丽思卡尔顿酒店公司的历史起源于波士顿丽思卡尔顿酒店。该波士顿地标的服务、餐饮及设施标准成为全球所有丽思卡尔顿酒店及度假村的基准。波士顿丽思卡尔顿酒店的传统始于有"酒店业者之王，王者之酒店业者"之称的著名酒店业者恺撒·里兹。通过他在巴黎里兹酒店和伦敦卡尔顿酒店的管理，他的服务与创新哲学重新定义了欧洲豪华酒店体验。波士顿丽思卡尔顿酒店以在酒店中创造豪华改革了美国的酒店业。丽思卡尔顿酒店公司的座右铭是"我们以绅士淑女的态度为绅士淑女服务"，而丽思卡尔顿全体工作人员的预期式服务态度正是最好的佐证。

About Ritz-Carlton

The Ritz-Carlton is a brand of luxury hotels and resorts with more than 70 properties located in major cities and resorts in 24 countries worldwide. The Ritz-Carlton Hotel Company LLC is now a wholly owned subsidiary of Marriott International, with currently 38,000 employees. The Ritz-Carlton headquarters are found in Chevy Chase, Maryland, a community along the border of Washington, D.C.

The history of The Ritz-Carlton Hotel Company, L.L.C. originates with The Ritz-Carlton, Boston. The standards of service, dining and facilities of this Boston landmark serve as a benchmark for all Ritz-Carlton hotels and resorts worldwide. The legacy of The Ritz-Carlton, Boston begins with the celebrated hotelier Cesar Ritz, the "king of hoteliers and hotelier to kings." His philosophy of service and innovations redefined the luxury hotel experience in Europe through his management of The Ritz Paris and The Carlton in London. The Ritz-Carlton, Boston revolutionized hospitality in America by creating luxury in a hotel setting. At The Ritz-Carlton Hotel Company, L.L.C., "We are Ladies and Gentlemen serving Ladies and Gentlemen." This motto exemplifies the anticipatory service provided by all staff members.

酒店概况

冲绳丽思卡尔顿度假酒店拥有 78 间设计现代的客房以及两间拥有露台和阳台的套房,将中国东海以及喜濑乡村俱乐部的美景尽收眼底。在深幽密林、壮丽的高尔夫球场、绚烂天空和浩瀚海洋等景观的环绕之中,冲绳丽思卡尔顿度假酒店突显了其独有的静谧和谐。

Overview

It owns 78 guestrooms with modern design and 2 suites with balconies and terraces where people could enjoy the beauty of East China Sea and the Kise Country Club. The hotel has shown its unique peace and harmony with the surrounding natural landscape like woods, sky and ocean, as well as the golf course.

Special 专题

Special 专题

酒店特色

琉球式城堡 Gusuku 是冲绳特有的古建筑，冲绳丽思卡尔顿度假酒店巧妙地将这一特色融入到建筑风格和景观设计之中，首里城色彩鲜明的红顶白墙和圣水坛，雅韵独特的冲绳式待客之道随处可见。

Feature

The Gusuku-style castles are peculiar ancient architectures to Okinawa, which are also been taken into the landscape design and architecture style of the Ritz Carlton Hotel. The unique Okinawa-style hospitality of red roofs and white walls could be found everywhere.

酒店配套

■ 餐饮

冲绳丽思卡尔顿度假酒店拥有精致雅韵的餐厅和酒吧，包括 Chura-Nuhji 意大利餐厅、以当地海鲜和世界闻名的冲绳牛肉为主的 Kise 日式铁板烧餐厅、拥有海边露台的 Gusuku 全天候餐厅，提供日式料理、冲绳特色及环球美食。

Services and Amenities

■ Dining

It is equipped with exquisite restaurants and bars, i.e. Chura-Nuhji Italian Restaurant, Kise Japanese Teppanyaki Restaurant famous for its Okinawa beef worldwide and local seafood, Gusuku All-Day Restaurant with seaside terrace, offering the food and drinks of Japanese style, Okinawa characteristic and global cuisines.

■ 水疗

丽思卡尔顿ESPA水疗中心拥有4间Spa理疗室、4间按摩室，以及户外凉亭，提供日本特色指压按摩或泰式按摩。水疗中心全部采用有机材质，深幽密林主题设计，搭配光影效果让客人仿佛置身于广袤苍穹下树影摇曳的原始森林。在理疗室内亦可欣赏到Yambaru原始森林的美景。水疗中心还设有美甲室、放松房、室内游泳池以及健身中心等。

■ Spa

ESPA Spa Center in the hotel has 4 spa psychotherapy rooms, 4 massage rooms and the outdoor arbor, providing the massage service of Japanese feature or Thai style. The center adopts the organic materials and jungle theme design with the effect of light and shadow to offer the clients experience of virgin forest. one could also enjoy the views of Yambaru native forest in the spa center. It also has the nail room, indoor swimming pools and fitness center etc.

Special 专题

Angsana Laguna Phuket
普吉岛乐古浪悦椿度假村

Keywords 关键词
- Contemporary 现代感
- Chic Retreats 回归自然
- Vibrant 生机
- Stylish 时尚感

品牌链接

悦椿是为追求品味、注重质感的现代旅客打造的酒店品牌。现代感十足并强调回归自然，悦椿为到此工作或休闲的宾客创造充满活力的体验。每一间悦椿酒店、度假村、Spa及精品店都将环保意识表现得淋漓尽致，同时力求将多元的个性色彩与亚洲文化遗产完美地融于一体。悦椿的每一处设施与各项服务，只为宾客尽情感受生命中的每一刻。

About Angsana

Angsana is a hotel brand that caters to the modern traveller seeking style and authenticity. Comprising contemporary and chic retreats, Angsana properties are designed to create and deliver vibrant enlivening experience for guests at work and at play. Each Angsana hotel, resort, spa and retail gallery exudes the spirit and conscience of its environment, while offering a strong sense of individuality infused with our Asian heritage. Facilities and services at all Angsana properties are focused on enabling guests to draw the most of every moment.

酒店地址：泰国普吉岛 Srisoonthorn 路四号
电　　话：+66 76 324 101

Address: 10 Moo 4 Srisoonthorn Road, Phuket, Thailand
Tel: +66 76 324 101

Special 专题

酒店概况

普吉岛乐古浪悦椿度假村位于泰国南部普吉岛西北方邦涛湾的海岸线上，从普吉国际机场和普吉镇出发只需20分钟车程。它是亚洲首个综合度假胜地——普吉岛乐古浪的一部分，普吉岛乐古浪拥有5座豪华酒店、6个棕榈摇曳的泻湖、400 000 m² 的绿荫、3 km 的原始海岸线和完善的餐饮娱乐设施。

Overview

Angsana Laguna Phuket is located on the shores of Bang Tao Bay in the northwest of Phuket Island in Southern Thailand, Angsana Laguna Phuket is just a 20-minute drive away from Phuket International Airport and from Phuket town. It is part of Laguna Phuket, Asia's first integrated resort, which consists of five luxury hotels, six palm-fringed lagoons, 600 acres of parkland, three kilometers of pristine beach and an outstanding array of recreational and dining facilities.

Special 专题

酒店特色

普吉岛乐古浪悦椿度假村的特色之处是为有活力的大都市旅行者设计的富有生机和时尚感的房间及设施。

Feature

Angsana Laguna Phuket features vibrant and stylish rooms and facilities designed for the dynamic cosmopolitan traveller.

酒店外观

普吉岛乐古浪悦椿度假村是融合了传统泰国建筑风格的当代杰作。拱峭的深拱形屋顶、自由无拘的空间布局、水道和泻湖在度假村内比比皆是。

Exterior

Angsana Laguna Phuket is contemporary in style with touches of traditional Thai architecture. Quintessential hallmarks such as steep arched roofs, free-flowing spaces, waterways and canals can be seen throughout the resort grounds. Modern elegance is conveyed in interior décor, via vivid-coloured fabrics, natural materials, stylish furnishings, fittings and abstract artwork.

| Special 专题

酒店室内

室内通过色彩鲜艳的材料、天然材料、时尚家具、配饰和抽象艺术品的装饰，散发出现代典雅气息。

Interior

Amazing views of lush tropical gardens, magnificent ocean vistas or the serene palm-fringed lagoon complement the resort's guest rooms, lofts, suites and private villa residences.

酒店配套

■ 餐饮

Market Place，以丰盛的自助早餐开启宾客在这里的全新一天，这个休闲用餐场所同样供应受欢迎的融合亚洲风味的美食、倍受人们喜爱的国际菜系及当地特色食物；Chao Lay，坐落在海滨，雅致的餐厅内及与白色沙滩仅有数步之遥的室外供应新鲜多汁的海鲜和传统的泰国美食；Bodega，这家餐厅供应品类繁多的葡萄酒和上好的意大利和地中海美食，宾客可以坐在别致的室内餐厅或搭设的露天用餐区享用所选美酒佳肴；树屋儿童餐馆，专为儿童设置，儿童餐馆菜单包括受儿童喜爱的、经健康再加工的汉堡、热狗和炸薯条等；海滩俱乐部，在悠闲的下午品着清凉鸡尾酒、享用俱乐部美食并在此招牌餐厅观望无与伦比的海景；Loy Krathong 酒吧，这里可以享用开胃酒、餐后酒、鸡尾酒、无酒精鸡尾酒或烈酒；池畔酒吧，这是度假村最受欢迎的酒吧，在这可点上一杯具有异国情调的鸡尾酒、新鲜果汁，品尝美味小吃和便餐。

Special 专题

Services and Amenities

■ Dining and Entertainment

Market Place: Start your day here with a sumptuous breakfast buffet. This casual dining venue also serves popular Asian-fusion cuisine, international favourites and local delights. Baan Talay: Located on the oceanfront, fresh succulent seafood and traditional Thai fare is served in the stylish interior of the restaurant or outdoors, steps away from the white sand beach. XANA Beach Club: Enjoy a leisurely afternoon of refreshing cocktails, club cuisine and unparalleled ocean views at this trendy dining destination. Loy Krathong Bar: Drop in for an aperitif, digestif, cocktail, mocktail or a stiff drink. In the evenings, your choice of beverage is served with a side of live entertainment. Bodega & Grill: Wine and dine at this restaurant offering an extensive selection of fine wine and the best of Italian and Mediterranean cuisine in an elegant indoor or al fresco setting. Poolside: Order exotic cocktails, fresh juices, tasty snacks and light meals at the resort's most popular watering hole. Kids Café: Specially catered to kids, the Kids Café menu features healthier renditions of favourites such as burgers, hotdogs and fries. Here, kids can also engage in fun activities before, during and after mealtime.

■ 会议

酒店共设有9个40~60 m² 不等的会议室和一个358 m² 的宴会厅。

■ Corporate Events

Meeting facilities: 9 meeting rooms from 40 sqm to 60sqm and a ballroom at 358 sqm.

■ 水疗中心

在屡获殊荣的悦椿 Spa 中通过亚洲灵感疗法刺激感官，焕发活力。融合了传统泰国疗法、西方技术以及新鲜植物萃取精华，专业护疗师每一个理疗过程都将令宾客身心松弛、焕然新生。

■ 健身设施

借助最先进的有氧运动和增强健身设备（瑜伽或普拉提斯课程或二者兼有），充满活力的课程可以实现宾客的锻炼目标。苍翠绿树和平静泻湖的壮丽景色为宾客的锻炼带来如沐春风般的舒适享受。

■ Spa

Set in Phuket's coveted resort destination, amidst the scenic Bang Tao Bay, Angsana Laguna Phuket features the island's premier family resort and exclusive spa haven. An idyllic setting by a tranquil lagoon, Angsana Spa Laguna Phuket offers a range of signature spa therapies catering to discerning spa lovers and families. Angsana Spa embraces the tropical garden spa concept and features a vibrant contemporary design and interior. It emphasises on the use of aromatherapy, therapeutic sense of touch and a fusion of techniques from the East and West. Its award-winning Spa Academies in Phuket, Thailand, Bintan in Indonesia and Lijiang in China are dedicated to the training of its professional and skilful therapists.

■ Fitness Facilities

Fulfil your exercise goals with a vigorous workout session utilising state-of-the-art cardio and strengthening equipment at the gym or with a Yoga or Pilates class or two. Magnificent views of lush greenery and the calm lagoon ensure your workout is a breeze.

Special 专题

■ 住宿

住宿体验既舒适又现代化：免费高速上网、咖啡/茶冲泡器具、iPod 扩展插口、液晶电视以及独特的悦椿系列用品。

■ 住宿

住宿体验既舒适又现代化：免费高速上网、咖啡/茶冲泡器具、iPod 扩展插口、液晶电视以及独特的悦椿系列用品。

■ Accommodation

Angsana Laguna Phuket has everything you want and more, which will make your stay comfortable and modern: Free use of in room high-speed internet access, tea and coffee-making facilities, iPod extension socket, LCD flat-screen TV and Angsana Laguna Phuket series.

Special 专题

Angsana Balaclava Mauritius
毛里求斯巴拉克拉瓦悦椿度假村

Keywords 关键词

Exotic Sensuality 异域风情
Ideal Hideaway 遁世之所
Mixed Culture 多元文化
Retreat Experience 休闲体验

品牌链接

悦椿是为追求品味、注重质感的现代旅客打造的酒店品牌。现代感十足并强调回归自然，悦椿为到此工作或休闲的宾客创造充满活力的体验。每一间悦椿酒店、度假村、Spa及精品店都将环保意识表现得淋漓尽致，同时力求将多元的个性色彩与亚洲文化遗产完美地融于一体。悦椿的每一处设施与各项服务，只为宾客尽情感受生命中的每一刻。

About Angsana

Angsana is a hotel brand that caters to the modern traveller seeking style and authenticity. Comprising contemporary and chic retreats, Angsana properties are designed to create and deliver vibrant enlivening experience for guests at work and at play. Each Angsana hotel, resort, spa and retail gallery exudes the spirit and conscience of its environment, while offering a strong sense of individuality infused with our Asian heritage. Facilities and services at all Angsana properties are focused on enabling guests to draw the most of every moment.

酒店地址：毛里求斯巴拉克拉瓦海龟湾
电　　话：+230 204 1888

Address: Turtle Bay, Balaclava, Mauritius
Tel: +230 204 1888

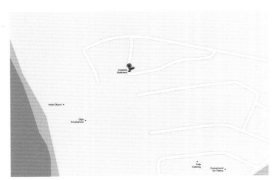

Special 专题

酒店开发管理

悦榕集团旗下的悦椿度假村至今营运超过10家酒店、超过40间Spa、以及超过40间精品店。悦椿是为追求品味、注重质感的现代旅客打造的酒店品牌。现代感十足并强调回归自然,悦椿为到此工作或休闲的宾客创造充满活力的体验。每一间悦椿酒店、度假村、Spa及精品店都将环保意识表现得淋漓尽致,同时力求将多元的个性色彩与亚洲文化遗产完美地融于一体。悦椿的每一处设施与各项服务,只为宾客尽情感受生命中的每一刻。

Development and Management

Managed by the Banyan Tree Group, Angsana Hotels and Resorts operates close to 10 resorts and hotels, over 40 spas, and 40 retail galleries. Angsana is a hotel brand that caters to the modern traveler seeking style and authenticity. Comprising contemporary and chic retreats, Angsana properties are designed to create and deliver vibrant enlivening experience for guests at work and at play. Each Angsana hotel, resort, spa and retail gallery exudes the spirit and conscience of its environment, while offering a strong sense of individuality infused with our Asian heritage. Facilities and services at all Angsana properties are focused on enabling guests to draw the most of every moment.

酒店概况

毛里求斯巴拉克拉瓦悦椿度假村位于毛里求斯西北部海岸线上的海龟湾，每晚面对壮观的日落，是一个远离尘嚣的伊甸园。它是为情侣和新婚夫妇设计的放松身心和享受私人空间的绿洲。

从普莱桑斯国际机场出发需要55分钟，如果从路易港出发只需15分钟即可到达首都之城，毛里求斯巴拉克拉瓦悦椿度假村是第一家5星级精品度假酒店，酒店拥有52间套房和别墅，设有亲密的原始海滩，并配备晚餐场所、以及顶级的娱乐和餐饮设施。

Overview

Nestled in the secluded area of Turtle Bay, which is in the North West coast of Mauritius, facing spectacular sunsets every evening, Angsana Balaclava Mauritius offers a hideaway ideal for travelers looking for privacy and relaxation. It is an oasis designed ideally for couples and honeymooners.

Located 55-minutes' drive from Plaisance International Airport and only 15 minutes' drive from Port Louis, the capital city, Angsana Balaclava Mauritius is the first 5-star boutique resort consisting of 52 suites and villa only, set on an intimate pristine beach, endowed with various dining venues, and an outstanding array of recreational and dining facilities.

Special 专题

Special 专题

酒店特色

巴拉克拉瓦悦椿度假村是该岛最顶级的精品度假村之一。在这里可以尽情陶醉于毛里求斯岛的自然景观和蝉声鸟叫,远离凡俗尘世,在天堂般的热带小岛和幽静的海滩上拥有世界级的休闲体验。

Feature

Angsana Balaclava Mauritius is the island's premier boutique resort – the latest 5-star resort on the Mauritian landscape. Let the world slip away as you sink into the natural sights and sounds of the island and experience a world class resort set in a paradisiacal tropical island and secluded beach.

酒店外观

座落在风景如画的亚热带毛里求斯海龟湾的巴拉克拉瓦悦椿度假村是岛内首屈一指的精品度假村。度假村的设计力求展现毛里求斯绝无仅有的特色——融合了东方、西方和非洲文化,其建筑和内部装潢独具特色,有覆盖茅草的屋顶、爬满棕榈藤的墙壁以及石面地板。度假村的规划非常谨慎细致,采取建筑与周围环境协调融洽的设计。度假村设计蕴含的理念是建造一所使身心和灵魂得到修复的天堂,以及夫妻和情侣的私密世外桃源。

Exterior

Set in subtropical Mauritius' picturesque Baie aux Tortues, or Turtle Bay, Angsana Balaclava is the island's premier boutique resort. Designed to reflect all that is uniquely Mauritian – the fusion of Oriental, Occidental and African cultural influences, the architecture and interior of the resort features thatched roofs, rattan wall coverings as well as stone flooring. In planning the resort, great care was taken to design buildings that are in harmony with their surroundings. The philosophy behind the resort is based on providing a haven for rejuvenation of the body, mind and soul, and an intimate hideaway for couples and honeymooners.

酒店配套

■ 餐饮

在这里可以心情愉悦地品尝大厨准备的、带有异国风情的亚洲美食、诱人的国际产品、正宗的东南亚食品和开胃的印度美食，客人们的选择甚多。毛里求斯巴拉克拉瓦悦椿度假村的经营理念旨在提供最美味的国际美食以及亚洲现代美食。为顾客定制畅游地点的主题餐饮，以便在度假胜地随时随地满足客人的需求。虽然健康美食有健身的功效，但酒店的理念是全身心的享受，而不是临床治疗。游客可以根据自己的喜好和需要，纵情于美食或养生菜肴。

Services and Amenities

■ Dining

Delight in the delectable cuisine prepared by our chefs — exotic Asian fusion fare, tantalizing international European offerings, authentic Creole food, and piquant Indian cuisine — where guests will be spoilt for choice. The Angsana Balaclava Mauritius concept aims to provide the most delicious international cuisines as well as contemporary Asian Fusion cuisine. Bespoke destination dining is provided to meet the needs of our guests any time and anywhere on the resort. While wellness cuisine is available to provide healthful benefits, the hotel's philosophy is holistic not clinical. Following their own desires and needs, guests can either indulge in gourmet dishes or ayurvedic cuisine.

■ **Rooms & Suites**

Picture perfect gardens adorned with tropical plants and palms create a haven for relaxation. As you wander the pathways to your room, the gentle breeze carries the sweet scent of local frangipanis and bougainvilleas lifting your mood and enhancing your senses. 52 suites and villa offer privacy and pampering designed to be your "home away from home". Bespoke in-room dinners create the perfect canvas for romance while the attentive staff are on hand to meet the guests' every need.

■ **套房和别墅**

完美如画的花园、摇曳的棕榈树、热带植物散发着馥郁气息,营造了一个令人身心放松的人间仙境。拾径而归,和煦微风携鸡蛋花和九重葛的甜美香气扑面而来,令人心醉的同时,神情不由一振。度假村的52间套房和别墅,尽享私密和尊贵感,如同"家外之家"。在房间内享用定制晚餐,完美的氛围尽显浪漫。酒店员工随时待命,为顾客奉上贴心服务,满足其所有需求。

■ 设施

悦椿Spa：毛里求斯巴拉克拉瓦悦椿度假村是享受顶级Spa之所在。在屡获殊荣的毛里求斯巴拉克拉瓦悦椿Spa，开启宾客的悦椿Spa之旅。诸多Spa护理、水疗体验和各类休闲活动，令人难以忘怀。悦椿Spa不仅提供令人神清气爽的护疗项目，建筑和室内装饰也同样卓尔不群。优雅的现代气息和毛里求斯多元文化传统完美结合，为宾客的体验锦上添花。

悦椿阁：悦椿阁以生活方式为导向，为游客提供当代亚洲度假村的体验和spa护理，以欣欣向荣的风格和活泼亮丽的色彩定义的商品系列为特色，反映出悦椿积极向上的生活态度。

■ Facilities

Angsana Spa: Angsana Balaclava Mauritius features the island's premier boutique resort and exclusive spa retreat. Award-winning Angsana Spa Balaclava Mauritius offers exclusive Angsana Spa Journeys, comprising a wide range of activities including spa treatments, hydrothermal treatments and leisure activities for a truly unforgettable spa journey. Complementing Angsana Spa's vitalizing treatments are stunning architecture and interior décor, which blend contemporary elegance with multi-cultural Mauritian heritage.

Angsana Gallery: A lifestyle-oriented gallery that offers visitors the option of bringing home a piece of the contemporary Asian resort and spa experience, Angsana Gallery features merchandise collection defined by energetic styles and electrifying colors, a reflection of Angsana's vivacious outlook on life.

Special 专题

Special 专题

Sheraton Yantai Golden Beach Resort

烟台金沙滩喜来登度假酒店

Keywords 关键词

- Seaside Resort 海滨度假
- Modern Luxury 现代奢华
- Culinary Experience 美食体验

酒店地址：中国山东省烟台市经济技术开发区（YEDA）海滨路88号
电　话：(86)(535) 611 9999

Address: 88 Haibin Road, Yantai Economic Development Area (YEDA) Yantai, Shandong, China
Phone: (86)(535) 611 9999

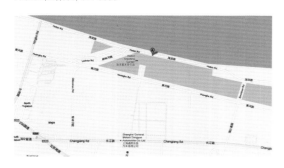

品牌链接

喜来登酒店与度假村集团（Sheraton Hotels and Resorts）是喜达屋（Starwood）酒店集团中最大的连锁旅馆品牌，而它也是集团中第二老的酒店品牌（最老牌的是威斯汀）。喜来登的酒店型态有许多种，从一般的商业旅馆到大型度假村都有；喜来登品牌一直力图维持高品质形象，在世界上的喜来登酒店有超过一半被当地机关评选为五星级酒店。喜来登酒店据点分布极广，遍布五大洲，从香港到斯里兰卡到埃及津巴布韦等国都可见其旅馆。喜来登总部在美国纽约的白原市。

喜达屋酒店都有良好的选址，主要分布在大城市和度假区。集团酒店选址的标准是：所在区域的发展史表明，该地区对提供全方位服务的豪华高档酒店有大量、持续增长的需求。作为酒店业豪华高档细分市场中最大的酒店集团，喜达屋酒店的规模有力地支持它的核心市场营销和预定系统。喜达屋酒店在把重点放在豪华高档细分市场同时，其各种品牌分别侧重于该市场中不同的二级市场。喜达屋酒店在赌场业也占据着重要的位置，它主要是通过Caesars品牌来经营此业务。喜达屋酒店为休闲度假旅游者提供着宾至如归（home-away-from-home）的服务。

About Sheraton Hotels and Resorts

Sheraton Hotels and Resorts is Starwood Hotels and Resorts Worldwide's largest and second oldest brand (Westin being the oldest). It occupies many types of hotel properties, ranging from general commercial hotels to large-scale resorts with high-quality image; more than half of Sheraton hotel worldwide are recognized as five-star hotels, with a widespread distribution across five continents i.e. Hong Kong, Sri Lanka, Zimbabwe etc. It headquarters in White Plains, New York.

Starwood hotels are located mainly in big cities or resort districts, in accordance with the site selection standards of group hotel that the selected districts own an increasing and great demand for full-serviced luxurious high-end hotels. As the greatest hotel group in the luxurious high-end market of hotel industry, Starwood holds its coral market promotion and booking system supported by its scale. It lays its emphasis on luxurious high-end market and its brands are targeted at the secondary markets. It also has a great achievement in casino industry, mainly represented by Caesars brand. Sheraton hotels address themselves to offer the visitors services of home-away-from-home.

酒店开发管理

2012年2月,烟台金沙滩喜来登度假酒店盛大开业。此为喜达屋酒店集团旗下首家酒店进驻山东烟台,诠释国际品牌新理念。

Development and Management

In February 2012, Sheraton Yantai Golden Beach Resort was opened to public. Which is the first hotel of Starwood Hotels & Resorts Worldwide, Inc that enter into Yantai, Shandong, explored the international brand concept.

酒店概况

烟台是中国东北部山东省的一座著名港口城市，地处渤海和莱州湾沿岸，四周环绕着美丽的海岛、湛蓝的海水和金色的沙滩。烟台金沙滩喜来登度假酒店位于烟台经济技术开发区（YEDA），地理位置优越，并与金沙滩隔路相望。酒店距市中心仅 15 km，距烟台莱山国际机场（YNT）仅 35 km。

Overview

Surrounded by islands, blue waters, and golden sands, Yantai—meaning "smoky tower"—is a popular port city located in China's northeastern Shandong Province, nestled along the Bohai Sea and Laizhou Bay. The Sheraton Yantai Golden Beach Resort is conveniently located across from Golden Beach, in the Yantai Economic and Technological Development Zone (YEDA). We are 15 kilometers from downtown, and just 35 kilometers from Yantai Laishan International Airport (YNT).

酒店特色

这里是远离都市喧嚣的绝佳休憩之地,先进完善的服务设施将令宾客身心放松、活力焕发、沟通无限!宾客可在设施齐全的水疗中心尽享舒适护理,或在天然纯净的沙滩之上漫步徜徉,亦可前往健身中心为健康充电,在网球场上挥洒汗水,或纵身于室内泳池中尽情畅游一番。独具特色的"随心连动@喜来登(SM),体验在微软®"位于酒店大堂区,设有现代化工作站、免费无线高速上网接入和一个可与亲友保持联系的舒适休闲区,宾客还可在提供全方位服务的商务中心利用高速上网接入跟进各项公务事宜。

Feature

A sophisticated retreat from the crowds and hectic pace of the city, the resort is well equipped to keep you refreshed, relaxed, and connected. Indulge in a treatment at our tranquil spa or simply stroll along the pristine beach. Stay in shape at our fitness center, play a match on the tennis court, and swim laps in our indoor pool. Out of town doesn't have to mean out of touch. The Link@Sheraton(SM) experienced with Microsoft® is a unique, comfortable lobby area featuring modern workstations, complimentary wireless High Speed Internet Access, and a lounge area for keeping up with friends and family. And stay in touch with the office at the full-service business centre, which also offers High Speed Internet Access.

酒店配套

■ 行政酒廊

入住喜来登行政客房的客人享有喜来登行政酒廊的特别使用权。行政酒廊是一处高档豪华的休闲空间，每日供应迎宾饮品、免费早餐、全天小吃、下午茶和晚间鸡尾酒。客人可来此与朋友和同事亲密聚会，或简单观看自己最喜爱的电视节目放松身心。行政级别客人还可享受免费的无线高速上网接入和一个独立的登记入住服务台。

■ 餐饮

无论是寻求一次美食之旅，一顿美味快餐，或是一边欣赏现场演出，一边啜饮晚间佳酿，酒店内的三间餐厅与一间酒吧都能让宾客流连忘返、乐不思蜀！

怡聚大堂吧，宾客可前来一边品尝美味咖啡和各式诱人甜点，一边欣赏壮观迷人的渤海湾美景，或在一天结束之际，遁入大堂吧的温馨氛围中，啜饮清爽鸡尾酒放松身心，并与亲朋好友畅谈交流；雅日本餐厅，选用来自日本的最上等食材，为宾客奉上各种生鱼片、寿司、铁板烧和其他美味选择，此外还提供各种清酒、烧酒和葡萄酒；盛宴西餐厅，荟集全球各地的顶级美食，餐厅配有开放式厨房和热情友好的厨师团队，将为宾客奉上令人难以抗拒的视觉及味觉盛宴；采悦轩中餐厅，供应原汁原味的粤菜和山东菜，是品尝一流美食的绝佳之地。

| Special 专题

Services and Amenities

■ Club Lounge

Sheraton Club Room guests have special access to the Sheraton Club Lounge. A relaxing, upscale space, the Club Lounge offers a welcome drink, complimentary breakfast daily, refreshments throughout the day, afternoon tea, and evening cocktails. Take advantage of the private Club Lounge where you can connect with friends, meet with your team, or simply relax by catching your favorite TV show. Club guests also enjoy complimentary wireless High Speed Internet Access and a separate registration counter.

■ Dining

Whether you're looking for a culinary journey, a quick bite, or an evening drink while enjoying live entertainment, the resort proudly offers three restaurants and a lounge. Taste cuisines from around the world as well as the best of Shandong Province. Using only the best and

freshest ingredients, the hotel is committed to serving every dish with innovative style and genuine hospitality for a truly unforgettable dining experience.

Connexions Lounge—While facing the stunning Bohai Bay, enjoy a cup of coffee and a great selection of cakes and pastries. Or relax at the end of the day with a refreshing cocktail and connect with family and friends in the welcoming ambiance of our lounge and bar. MIYABI—Using only the finest ingredients from Japan, take you on a culinary journey with a variety of sashimi, sushi, teppanyaki, and other delicious options. Offer such beverages as sake, shochu, and wines. FEAST—Feast welcomes you with the finest gourmet food from around the world. Here provides guests with an irresistible visual and culinary spread with open kitchens and friendly chefs that encourage an interactive dining experience. YUE—Gather with friends and family for good conversation and great food at Yue—the Chinese specialty restaurant that offers authentic Cantonese and Shandong cuisines in a relaxing atmosphere.

■ 会议与活动

烟台金沙滩喜来登度假酒店提供 2 300 m² 的多功能空间与卓越一流的服务设施，包括面积宽达 923 m²，可容纳 600 名客人的华新红宝石宴会厅，以及 7 间面积在 42~144 m² 不等的多功能厅，可适应不同的会议需要。现代化的视听设备，包括先进的演示技术、专业的音响和照明系统，以及高速上网接入均一应俱全。

酒店同时还是一处梦幻般的婚礼场地，酒店的婚礼专家将全程为客人打理每一细节，从而确保客人的活动完美无瑕。无论举办何种活动，经验丰富的活动专家都将随时助上一臂之力。

■ Meetings & Events

With 2,300 square meters of function space complemented by first-class service, Hua Xin Ruby Ballroom spans 923 square meters and can hold up to 600 guests. An additional seven multifunction rooms ranging in size from 42 to 144 square meters ensure you will find a perfect space. Modern audiovisual equipment, including advanced presentation technology and professional sound and lighting, and High Speed Internet Access are readily available.

The resort is a dreamlike wedding venue. The special wedding planner will take care of every detail, from your first meeting until your last guest leaves, ensuring that your event is flawless. For any event, experienced and helpful specialists will assist you every step of the way.

Special 专题

■ 客房

烟台金沙滩喜来登度假酒店 363 间舒适典雅的客房及套房均拥有充足的自然采光，并可欣赏到壮丽的海洋和沙滩美景。每间客房均配备享有盛誉的喜来登甜梦之床™，可确保宾客安享一夜好眠。一道玻璃幕墙将卧室与精美的大理石浴室分隔开来，浴室内配有深浸泡式浴缸、带六喷头雨林花洒的淋浴间、放置于中央的悬浮式面盆以及一系列卫浴用品。

■ Rooms

The Sheraton Yantai Golden Beach Resort offers some of the most elegant and stylish accommodations in the city. 363 tastefully appointed guest rooms and suites are suffused with natural light and feature magnificent sea and beach views. The famous Sheraton Sweet Sleeper™ Bed which guarantees a fabulous night's sleep. Glass walls separate the bedroom from the exquisite marble bathroom, which features a deep-soaking bathtub, a shower with a six-jet rainforest showerhead, a central floating washbasin, and a variety of bath amenities.

Sheraton Bali Kuta Resort
巴厘岛库塔喜来登度假酒店

Keywords 关键词
- Chain Brands 连锁品牌
- Beach Resort 滨海度假
- Beach Walkway 海滩步道
- Tropical Temptation 热带风情

酒店地址：印度尼西亚巴厘岛库塔库塔海滩
电　　话：(62)(361) 846 5555

Address: Jalan Pantai Kuta, Kuta, Bali, Indonesia
Tel: (62)(361) 846 5555

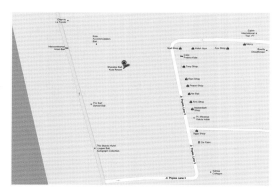

品牌链接

喜来登酒店与度假村集团（Sheraton Hotels and Resorts）是喜达屋（Starwood）酒店集团中最大的连锁旅馆品牌，而它也是集团中第二老的酒店品牌（最老牌的是威斯汀）。今日的喜来登品牌是在1937年出现的，当时两位企业家Ernest Henderson以及Robert Moore在马萨诸塞州的斯普林菲尔德成立了第一家喜来登酒店。1945年喜来登成为第一家在纽约证券交易所挂牌上市的连锁酒店集团。1995年福朋喜来登品牌成立，喜来登希望以合理的价格提供全方位的服务；当时很多规模较小的喜来登酒店都被改名为福朋喜来登。1998年喜达屋集团以高于希尔顿的出价收购了喜来登品牌。在喜达屋的管理领导下，喜来登开始创建更多的酒店以扩大其品牌影响力。

About Sheraton

Sheraton Hotels and Resorts is the largest hotel chain brand of the Starwood Hotels and Resorts Worldwide Inc. and second oldest brand (Westin being the oldest). The origins of the sheraton dated back to 1937 when Ernest Henderson and Robert Moore acquired their first hotel in Springfield, Massachusetts. In 1945, it was the first hotel chain to be listed on the New York Stock Exchange. In 1995, Sheraton introduced a new, mid-scale hotel brand Four Points instead of Sheraton Hotels, Which provides a comprehensive service at a reasonable price. In 1998, Starwood Hotels & Resorts Worldwide, Inc. acquired Sheraton, outbidding Hilton. Under Starwood's leadership, Sheraton has begun renovating many existing hotels and expanding the brand's footprint.

Special 专题

酒店概况

巴厘岛素有"众神之岛"之称，坐落于爪哇岛和龙目岛之间，其浓郁的异域风情和神秘色彩必将为旅客带来一次真正独特的旅行体验。巴厘岛库塔喜来登度假酒店俯瞰着岛上著名的库塔海滩和惹班底库塔路沿岸的巴厘海峡，隶属于海滩步道的一部分；海滩步道上拥有萨希德库塔时尚度假村、购物中心、餐厅、酒吧、夜生活场所和XXI电影城。努拉莱国际机场，也称为登巴萨国际机场（DPS），距度假酒店仅15分钟车程。

Overview

Often called the "Island of the Gods" and nestled between Java and Lombok, exotic and mystical Bali promises a truly unique experience. Overlooking the island's famed Kuta Beach and the Bali Strait along Jalan Pantai Kuta, the Sheraton Bali Kuta Resort is part of Beachwalk: A Sahid Kuta Lifestyle Resort, a lifestyle center complete with shopping, restaurants, bars, nightlife, and an XXI Cineplex. Ngurah Rai International Airport, also known as Denpasar International Airport (DPS), is only a 15-minute drive away.

酒店特色

无论是忙于公务事宜还是享受休闲放松,酒店都可为房客提供多重精彩选择。酒店大堂内的"随心连动@喜来登(SM),体验在微软®"为房客开辟了一处与新知故交聚会畅谈的理想之地。也可利用酒店内的电脑终端体验网上冲浪的无穷乐趣,或带上自己的笔记本电脑享受这里的免费无线高速上网接入。

喜来登与Core® Performance的培训专家携手合作,为客人打造一项全面综合的健身康体计划,以帮助其在旅途中随时保持健身日程。酒店的健身中心面积达160 m^2,可提供毛巾、饮水站与健身指导。有氧健身器材配有独立的电视屏幕,而力量与核心训练区则配有一台42英寸的液晶纯平电视。此外,度假酒店内还设有一条慢跑路线。

无论是在自由形泳池还是儿童泳池中,每个人都有一方属于自己的嬉水空间。酒店可提供躺椅和遮阳伞,以及食品与饮料服务。邻近的海滩步道还设有水上乐园和儿童保育中心。

Feature

Whether the clients want to stay busy or relax, the resort has plenty of options for them. In the lobby, which is designed so they can socialize with friends new and old, is the Link@Sheraton(SM) experienced with Microsoft®. Feel free to surf the web on the terminals, or bring the laptop for complimentary wireless High Speed Internet Access.

Sheraton works in partnership with training experts at Core® Performance to offer guests a comprehensive health and fitness program that helps the clients maintain their routine while on the road. The 160-square-meter fitness center provides towels, a water station, and instruction posters. Cardio equipment features individual television screens, and the strength/core area has a 42" LCD flat screen television. Furthermore, a jogging route is available in the resort.

Everyone will have a place to splash around, whether in the free-form pool or the kids' pool. Lounge chairs and umbrellas are available, as is food and drink service. A water park and a child-care center operate out of the adjacent Beachwalk.

Special 专题

酒店配套

■ 餐饮

度假酒店的餐饮选择体现了巴厘岛丰富的天然食材和香料。酒店的"点亮餐盘"计划是 Core™ Performance 喜来登健身计划的一部分,其中包括营养颜色代码。菜单上的所有菜品均含有各色蔬果中的主要营养成分,旨在确保适当的饮食平衡。

酒店的盛宴全日制餐厅设有开放式厨房,可为来客打造互动式的完美用餐体验。自助餐和零点菜肴丰盛齐全,为客人呈现色香味俱全的新鲜美食。客人还可在海滩步道旁的室外露台上享用大餐。

现代感十足的班妮餐厅是一家正宗地道的意大利餐厅,其氛围轻松活跃,适合于举家同乐。餐厅位于池畔,可从上层露台饱览美不胜收的迷人景致。餐厅对传统的意大利菜肴进行了现代化的演绎,并采用当地食材和从巴厘岛水域中捕获或进口的新鲜海鲜。

在舒适宜人的大堂吧中,客人可品茗一口香浓咖啡,或啜饮一杯清爽佳酿,彻底放松身心。也可在欣赏壮观海景的同时,尽情品尝各式小吃。

■ 会议

巴厘岛库塔喜来登度假酒店坐拥库塔海滩的壮丽美景,将为来客打造真正令人难忘的活动体验。酒店的九间会议室总面积超过 1 600 m²,均以 Pawayangan 的著名人物来命名,反映了印尼浓厚的文化底蕴。700 m² 的多功能豪华宴会厅可被分成几处较小的空间。

全酒店上下各具特色的场地布置十分适合于举办露天聚会、主题派对及其他特别活动。酒店将不断致力于提供先进一流的技术与现代化设施,以激发与会者的无限创意灵感,而无线高速上网接入更可让客人与外界随时保持联络。

Services and Amenities

■ Dining

The resort's dining options reflect Bali's abundance of natural ingredients and spices. The Color Your Plate initiative incorporates the Nutrition Color Code, part of Sheraton Fitness Programmed by Core™ Performance. All menu items feature key nutrients from colorful fruits and vegetables to ensure a proper dietary balance.

At Feast, the all-day dining restaurant, open kitchens encourage an interactive dining experience. Only fresh, flavorful produce and ingredients are used in the expansive buffet and à la carte dishes. Guests may enjoy dining on the outdoor patio along Beachwalk.

Authentically Italian, Bene is a light-hearted, fun-filled modern trattoria that the whole family will enjoy. Its poolside location provides a breathtaking view from the upper terrace. Traditional Italian meals take on a modern twist, using local produce and fresh seafood either caught in Bali's waters or imported.

In the comfort of The Lounge, relax with a delicious cocktail, glass of wine, or espresso. Share a variety of tapas while taking in the stunning ocean view.

■ Meetings

Overlooking Kuta Beach with stunning ocean views, every event at Sheraton Bali Kuta Resort will be truly memorable. Totaling more than 1,600 square meters, nine meeting spaces are named after famous characters of Pawayangan to reflect Indonesia's rich culture and heritage. A versatile 700-square-meter Grand Ballroom can be divided into smaller areas.

Open-air functions, themed parties, and special events can be arranged at a variety of unique locations throughout the resort. The ongoing commitment to provide state-of-the-art technology and modern facilities invites a meeting of the minds, and wireless High Speed Internet Access allows guests to stay connected.

■ 客房

酒店拥有203间设备先进的客房与套房,糅合了时尚的巴厘岛风情与奢华格调。走出宽敞的私人阳台,或穿过大门步入热带社交庭院,房客即可欣赏到库塔海滩和巴厘海峡的迷人美景。

每件家具都反映了一种归属感,珍珠母贝壳图案体现的是美丽的印度洋风情。现代便捷设施包括无线高速上网接入、LED或液晶纯平电视、可与iPod连接的多媒体系统以及国际通用插座。茶点中心或免费瓶装水可为房客补充能量。

漫长一天结束后,房客可在铺有豪华床垫、上好棉布床单、舒适羽绒被和鹅绒枕的喜来登甜梦之床™中尽享安眠。宽敞的水疗浴室美观大方,配有独立浴缸和雨林淋浴间。

■ Rooms

Contemporary Balinese taste and opulence seamlessly blend with state-of-the-art technology in the 203 rooms and suites. Step out onto the spacious private balcony for enchanting views of Kuta Beach and Bali Strait, or walk through the door to a tropical social courtyard.

Every piece of hand-selected furniture reflects a sense of belonging, including mother-of-pearl shell motifs that reflect the beautiful Indian Ocean. Modern conveniences include wireless High Speed Internet Access, an LED or LCD flat screen television, a multimedia system with iPod© connectivity, and an international electrical outlet. Relax with the refreshment center or complimentary bottled water.

At the end of a long day, rest comfortably in the inviting Sheraton Sweet Sleeper™ Bed with its plush mattress, fine cotton sheets, cozy duvet, and down pillows. Spacious spa-inspired bathrooms are beautifully appointed with freestanding bathtubs and rainforest showers.

Resort Hotel 度假酒店

The St. Regis Bal Harbour Resort | 巴尔港瑞吉度假酒店

Keywords 关键词
- Business and Leisure 商务休闲
- Bespoke Service 定制服务
- Seashore Landscape 海滨风光

酒店地址：美国佛罗里达迈阿密海滩巴尔港科林斯大道9703
电　　话：(305) 993-3300

Address: 9703 Collins Avenue Bal Harbor - Miami Beach, Florida. United State
Tel: (305) 993-3300

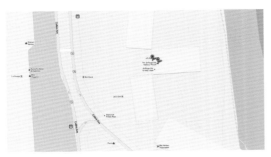

品牌链接

瑞吉酒店是喜达屋旗下的奢侈品牌，创于1999年。瑞吉酒店面向高端商务和休闲旅客提供完美无瑕的定制服务，是世界上最高档酒店的标志。作为优雅与精致的最佳典范，瑞吉酒店及度假村对卓越的追求可谓精益求精。确保每次住宿都能达到客人的最高标准，满足其高贵典雅的独特品味。酒店还通过瑞吉迷（St. Regis Aficionado(SM)）诚邀客人利用独享特权表达对瑞吉的热爱之情，其中包括西斯廷教堂私人游、品尝彼特鲁庄园稀有精选佳酿的难得机会或有私人主厨和品酒师相伴的乡村滑雪之旅。

而今，瑞吉酒店及度假村已经遍布全球的多个地点。如伦敦、纽约、新加坡、巴厘岛——每家酒店都能让宾客领略到当地的迷人风情和独特风韵。此外，每家酒店还一如既往地承袭了首家瑞吉酒店所确立的豪华与精致标准，并在不断发展中又纳入五种与众不同的设计风格：都市庄园、玻璃屋、精致半球、旅程终点和天堂乐园。每种设计风格均以独特手法完美再现了瑞吉品牌的丰富精髓与悠久传统，例如宽大的楼梯、熠熠闪光的吊灯、雅致的图书馆、琳琅满目的酒窖、标志性壁画和青铜正面装饰。

About St. Regis

St. Regis is Starwood's main luxury brand, launched in 1999. It is the sign of the most upscale hotels in the world providing flawless and bespoke service to high-end leisure and business travelers. As exemplars of elegance and refinement, St. Regis hotels and resorts are uncompromising in their pursuit of excellence. Every stay is commissioned to meet guests' highest standards and refined to express the subtlety of their unique tastes. Through St. Regis Aficionado(SM), guests of the hotel are invited to indulge their passions with exclusive privileges such as a private tour of the Sistine Chapel, the opportunity to sample a rare vertical selection of Chateau Petrus, or a back country ski trip with a personal chef and sommelier.

Today St. Regis hotels and resorts can be found across the globe. London, New York, Singapore, Bali - each is an entrance into a captivating world of seduction and a unique expression of its location. The standard of opulence and sophistication established by the original St. Regis is honored in every address, but has evolved to include five distinct design interpretations: Metropolitan Manor, Glass House, Hemispheres, Journey's End and Paradise Found. In each, the essence of the brand and its rich traditions is brought to life through signature features such as grand staircases, glittering chandeliers, handsome libraries, vast wine vaults, iconic murals and bronze façades.

Resort Hotel 度假酒店

Resort Hotel 度假酒店

酒店概况

巴尔港瑞吉度假酒店及其公寓位于迈阿密最负盛名的南佛罗里达州海滩。度假酒店拥有251间典雅的客房与套房，并特设269间私人公寓。度假酒店由Sieger Suarez Architectural Partnership公司打造，设计精湛，细节完美。这间独特的度假酒店对面是著名的巴尔港商店，距迈阿密与南部海滩仅几分钟的路程。度假酒店为宾客奉上瑞吉特色管家服务，还开设Remède水疗中心，可享受顶级的水疗和健身中心的先进设施，更有高级餐厅、儿童看护、池畔小屋与沙滩服务可供选择。

Overview

The St. Regis Bal Harbour Resort is located on the most famous beach in south Florida, with 251 elegant guestrooms and suites, and 269 personal apartments. It is constructed by Sieger Suarez Architectural Partnership, with exquisite design and perfect details. The hotel is located facing the legendary Bal Harbour Shop and several minutes' way to the dynamic Miami and South Beach. It offers the butler service with Regis characteristic, and the top-tier spa and fitness with the advanced facilities in Remède Center. The customers could enjoy the high-end restaurants, child care, and pool cottage, beach service here.

| Resort Hotel 度假酒店

酒店特色

巴尔港瑞吉度假酒店致力于为客户提供定制的体验和服务,其标志性的管家服务和礼宾服务可以满足客户的各个需求,包括安排到南海滩以及之外的交通。瑞吉海滩俱乐部拥有两个壮观的游泳池——面向成人的宁静游泳池和面向家庭的度假酒店游泳池,还提供私人小屋出租服务,客人可体验这里郁郁葱葱的林木和私人海滩。另外,这里具有国家最先进的健身套房以及约 1 300 m^2(14 000 平方英尺)的 Remède Spa 供客户舒缓身心。儿童俱乐部为年幼的客人提供一系列精彩的活动。

> Feature

The St. Regis Bal Harbor Resort is dedicated to providing a bespoke experience, the iconic St. Regis Butler Service and Concierge Service are always on hand to anticipate guests' every need, including arranging transportation to South Beach and beyond. Meanwhile, The St. Regis Beach Club comprises two spectacular pools—the adult-oriented Tranquility Pool and family-friendly Resort Pool—and private cabana rentals, inviting guests to enjoy nine lush acres and private beach access. Further, the state-of-the-art Fitness Suite keeps guests in shape while the 14,000-square-foot Remède Spa soothes the body and the mind. For youngest guests, Kids' Club offers an exhilarating variety of activities.

Resort Hotel 度假酒店

酒店配套

■ 餐饮

酒店拥有 3 个餐厅，佐以精美的室内和户外环境，为客户提供难忘的美食享受。客人以 Atlantico 的丰盛早餐作为一天的开始，伴随着阳光体验 Fresco Beach Bar and Grill 所提供的海边和池畔用餐。午餐和晚餐可以选择 J & G Grill，这是酒店的招牌之一，灵感来自国际名厨，由 de Cuisine Richard Gras 厨师创立。夜晚可以在瑞吉酒吧和葡萄酒窖参加香槟仪式，欣赏现场爵士乐。瑞吉管家和私人厨师随时应客户要求在个人客房或海滨白昼别墅为其准备盛宴。

Services and Amenities

■ Dining

Harmonizing epicurean offerings with exquisite indoor or al fresco surroundings, three restaurants and signature bar inspire unforgettable culinary experiences. Enjoy a sumptuous breakfast in Atlantico, before a day in the sun with casual oceanfront and poolside dining at Fresco Beach Bar and Grill. Savor lunch or dinner at the hotel's signature J&G Grill, inspired by celebrated international chef Jean-Georges Vongerichten, and brought to life by Chef de Cuisine Richard Gras. Start evening with the champagne-sabering ritual and live jazz and bossa nova music in the St. Regis Bar and Wine Vault. At any hour of the day, St. Regis Butler and personal chef are on call to provide a memorable private meal in guest room or Oceanfront Day Villa.

Resort Hotel 度假酒店

■ 会议与活动

巴尔港瑞吉度假酒店设置了约 1 040 m² (约 11 200 平方英尺) 装潢精美的室内会议和活动空间, 以及各种迷人的海滨户外场地。从华丽的首映礼, 到创意主题活动、贴心的商务会议、婚礼等, 酒店均提供餐饮和经验丰富的活动专业人员, 参与每一个细节。另外, 最大的视听设施和国家最先进的商务中心仅几步之遥, 确保满足所有的技术需求。

■ Events

The St. Regis Bal Harbor Resort Offers approximately 11,200 square feet of exquisitely appointed indoor meeting and event space, along with a variety of stunning oceanside outdoor venues. From opulent galas to creatively themed events to intimate business meetings to legendary weddings, the hotel provides custom catering and experienced event professionals, who attend to every detail. Additionally, premier audiovisual amenities and a state-of-the-art Business Center, just steps away, ensure all technological needs are easily met.

Resort Hotel 度假酒店

■ 客房

酒店拥有251间典雅的客房与套房。客房内配有瑞吉名牌床垫和包裹在400针的Pratesi床单内的带古董镜子的床头板,并搭配有羽绒枕和鸭绒被。酒店客房采用领先技术,提供高速上网端口,带智能控制管家服务技术的床头控制板以及娱乐设施,如带蓝牙播放机的46寸三星LED电视、Ipod基座和环绕声系统。大型的步入式衣橱和镜像抽屉提供了充足的存储空间。

■ Guestrooms

The St. Regis Bal Harbor Resort owns 251 elegant guestrooms and suites in which signature pillowtop St. Regis beds framed by antique mirror headboards, are wrapped in crisp 400-thread-count Pratesi bed linens and topped by down pillows and comforters. Premier technology, including High Speed Internet Access and a bedside control panel featuring intelligent control butler service technology, ensures intuitive living while leading entertainment amenities comprise a 46-inch Samsung LED HDTV with Blu-ray player, iPod docking station, and surround sound system. Large walk-in closets, complete with mirrored drawers, offer ample storage space.

Resort Hotel 度假酒店

Resort Hotel 度假酒店

Jumeirah Port Soller Hotel & Spa | 卓美亚 Port Soller 水疗酒店

Keywords 关键词
- Artistic Connotation 艺术内涵
- Local Specialty 地方特色
- Customer Oriented 客户至上
- Wellness and Fitness 健身休闲

品牌链接
卓美亚酒店堪称世界上最奢华、最具创新意识的酒店，已荣获无数国际旅游奖项。集团始建于1997年，志存高远，立志通过打造世界一流的奢华酒店及度假酒店，成为行业领袖。在这些辉煌成就的基础上，卓美亚集团 (Jumeirah Group) 于2004年成为迪拜控股旗下一员，在这艘汇聚了迪拜众多顶尖企业和项目的航母上，卓美亚驶向了成长与发展的新旅程。

About Jumeirah
Jumeirah Hotels & Resorts are regarded as among the most luxurious and innovative in the world and have won numerous international travel and tourism awards. The company was founded in 1997 with the aim to become a hospitality industry leader through establishing a world class portfolio of luxury hotels and resorts. Building on this success, in 2004 Jumeirah Group became a member of Dubai Holding – a collection of leading Dubai-based businesses and projects – in line with a new phase of growth and development for the Group.

酒店地址：西班牙马洛卡岛索尔拉港07108号 Calle Bégica s/n
电　　话：+34 971 637 888

Address: Calle Bégica s/n, 07108 Puerto de Sóller, Mallorca, Spain
Tel: +34 971 637 888

Resort Hotel 度假酒店

酒店概况

酒店位于马洛卡岛的西北面海岸，四周风景如画，毫无任何人工破坏的痕迹。酒店总占地 18 000 m²，由 11 栋彼此独立的建筑组成，共设有 120 间客房。酒店设计即考虑整体的建筑方案使其很好地融入到周围纯天然的山景之中。

酒店建在一座陡峭的悬崖上，仿佛漂浮在蓝天碧海之间。从这里可以俯瞰索尔拉港口的一座小渔村，四周被崇山峻岭所环抱，这里已经被联合国教科文组织列为世界历史遗产名录。酒店距离 Palma 国际机场仅 35 km。

Overview

Located on the picturesque and unspoilt Northwest coast of Mallorca, Jumeirah Port Soller Hotel & Spa brings a new standard of luxury to the Spanish island without any trace of artificial damage. The hotel has 120 rooms spread over 11 separate buildings covering 18,000 sq metres, all connected by garden walks and internal passageways. This intimate hotel ensures that guests will feel part of the pure natural mountain landscape.

Built on a cliff overlooking the fishing village of Port Soller and surrounded by the Tramuntana Mountain range, a UNESCO Heritage site, it appears to float between the sky and the sea. The hotel is 35 kms away from Palma International Airport.

Resort Hotel 度假酒店

酒店特色

卓美亚 Port Soller 水疗酒店得天独厚的地理位置令其成为自然爱好者的天堂，他们可以徒步或骑山地自行车探索这里将近 200 km 的小道，又或者在岛上 26 座高尔夫球场中任选一座享受高尔夫的乐趣。

Feature

The advantageous geographic position of Jumeirah Port Soller Hotel & Spa has enabled it become the destination of nature lovers. They could walk or ride on bicycles to explore the 200 km path or randomly pick up a golf course among the 26 on the island to enjoy the pleasure.

Resort Hotel 度假酒店

Resort Hotel 度假酒店

酒店外观

卓美亚 Port Soller 水疗酒店建在索尔拉港西北角的一座悬崖上，俯瞰着壮丽的索尔拉湾。酒店分为 11 座低层建筑，均与当地景观巧妙地融为一体。

Exterior

Jumeirah Port Soller Hotel & Spa is situated on a cliff in the northwest corner of Port Soller, watching over the beautiful Soller Bay. The hotel consists of 11 low buildings, all blending beautifully with the local landscape.

酒店室内

酒店内部装饰经过精心设计，营造出现代而舒适的氛围。不同色调的米色和大地色系加上橄榄木和松木等当地木材、当地手工艺品及地方特色建筑、宽敞的房间，及俯瞰高山、大海以及港口的落地窗，正是这些元素为客人带来了无与伦比的体验。

卓美亚 Port Soller 水疗酒店的艺术收藏也是经过精挑细选的。这里的收藏被称为"艺术体验"，由 400 件当地的艺术家创作的作品组成。艺术品的灵感主要来自于酒店周围岛屿的自然环境，艺术品的种类包括绘画、水彩、拼贴、照片、石雕、陶瓷、玻璃作品以及绳索艺术等。

Interior

The interior of the hotel has been deliberately decorated to present modern and comfortable atmosphere. Olive wood and pine of different beige colours and earth tone, local artifacts and distinctive buildings, spacious rooms and French windows to look over mountains, sea and the port, all the elements together offer fantastic experience for their guests.

The art collection at Jumeirah Port Soller Hotel & Spa has specifically been created and managed. The collection, called THE ART EXPERIENCE, is made of 400 artworks produced by a dozen local artists. Inspiration for the artworks came from the surroundings and Nature has been the artists' muse. The collection includes paintings, watercolours, collages, photographs, sculpted stones, ceramics, glasswork and even ropes.

酒店配套

■ 美食休闲

酒店开设的两家餐厅带来了非凡的美食体验,其中的 Es Finland Cap Rig 餐厅绝对是马洛卡岛不容错过的美食享受,其创新菜式具有强烈的本地特色,完全采用最新鲜的时令食材精心制作而成。

■ 泰丽丝水疗中心

泰丽丝是马洛卡岛上的唯一一家康体健身场馆,共分为10个场地宽大的康体健身区域,如澡堂套系、高级健身房以及最新开发的"泰诺健"健身设备。场馆内的私人健身教练会随时对客人进行指导和评估。这座健身场馆总面积超过 2 200 m²,从各个角度均可以看到酒店外面壮丽的风光,从室外的水疗泳池还可以俯瞰山谷的风光。

■ 户外活动

在室外三座芬芳的花园里设有三个游泳池,其中两个全年进行加热,在这里游客不仅可以饱览地中海沿岸的独特美景以及松林保护区内的壮丽景色,还可以进行鸡尾酒舞会或者野餐,放松身心。这一区域可供游客进行多项室外活动,大多都与山有关,如徒步旅行、登山、山地自行车和攀岩等。在日间还可以租赁私人游艇出游,欣赏周围绿松石般的海水和小湾,或者钓鱼、潜水等。从这里进行短暂航行之后就可到达梅诺卡岛和伊比莎岛。

Resort Hotel 度假酒店

Services and Amenities

■ Dining and Entertainment

The two restaurants run by the hotel bring extraordinary culinary experience to their guests while Es Finland Cap Rig Restaurant provides not-to-be-missed culinary enjoyment. The restaurant uses fresh local seasonal materials to deliberately create dishes with strong local flavour.

■ Talise Spa

Jumeirah's signature Talise Spa is set to become the island's most exclusive wellness and fitness venue. It comprises of 10 spacious treatment suites including a Hammam Suite, a state-of-the-art gym with the latest Technogym equipment, while personal trainers are available for assessments and guidance. With over 2,200 sq meters and stunning views from every corner, it features its own outdoor Hydropool overlooking the valley.

■ Outdoor Activities

Amongst the hotel's scented gardens, three swimming pools (two of them heated all year round) and several vantage points provide uninterrupted views of the Mediterranean and the protected pine forest surrounding the property, creating ideal spaces for outdoor relaxation and fantastic terraces for cocktails and dining al fresco. The area offers a wealth of outdoor activities, mostly connected to the stunning mountains which are renowned for trekking, hiking, mountain biking and climbing. Private yachts are available to hire for day excursions to the turquoise water beaches and nearby coves, as well as fishing, snorkelling and scuba diving. The islands of Menorca and Ibiza are also within short sailing distance.

Resort Hotel 度假酒店

图书在版编目（CIP）数据

酒店⁺1　滨海度假：汉英对照 / 佳图文化 主编 . -- 北京：中国林业出版社，2013.7

ISBN 978-7-5038-7075-0

Ⅰ . ①酒… Ⅱ . ①佳… Ⅲ . ①建筑设计—图集　②景观设计—图集　Ⅳ . ① TU206　② TU986.2-64

中国版本图书馆CIP数据核字（2013）第079563号

策　　划：王　志
主　　编：佳图文化

中国林业出版社·建筑与家居图书出版中心
责任编辑：李　顺　唐　杨
出版咨询：（010）83223051

出　版：中国林业出版社（100009　北京西城区德内大街刘海胡同7号）
网　站：http://lycb.forestry.gov.cn/
印　刷：利丰雅高印刷（深圳）有限公司
发　行：中国林业出版社
电　话：（010）83224477
版　次：2013年7月第1版
印　次：2013年7月第1次
开　本：889mm×1194mm　1/16
印　张：18
字　数：200千字
定　价：328.00元（USD 60.00）